Immunocytochemistry

Richard W. Burry

Immunocytochemistry

A Practical Guide for Biomedical Research

 Springer

Richard W. Burry
College of Medicine & Public Health
Ohio State University
333 West 10th Avenue
Columbus, OH 43210-1239
USA
burry.1@osu.edu

DOI 10.1007/978-1-4419-1304-3
Springer New York Dordrecht Heidelberg London

Printed on acid-free paper

Springer is part of Springer Science+Business Media (www.springer.com/mycopy)

To Yvonne, my best friend, my wife, and my technical editor, for her love and unwavering support of this project. And to my parents, for lighting the fire in me as a child by giving me a microscope.

Acknowledgments

Thanks to my many colleagues in the Histochemical Society and at The Ohio State University for their discussions that led to the concept of this book. Specifically, to Elizabeth Unger for reading the manuscript at various stages and whose ideas were invaluable and gave me a different understanding of immunocytochemistry; to Paul Robinson for the insight his years of experience gave; to Amy Tovar for centering my ideas; to John Gensel for help with the case studies; to Vidya Kondadasula for ideas early in the project; Mary Jo Burkhard for editing, and to Georgia Bishop for help with organization. Thanks to Ping Wei and Wenmin Lai for the technical help with sectioning. Special thanks to Yvonne Burry for the hours of reading and editing the manuscript. Thanks to Carol Larimer for her editing expertise. Thanks to Stephanie Jakob of Springer for her encouragement and support of this project.

Contents

1 Introduction 1
What Is Immunocytochemistry? 1
What Can Immunocytochemistry Tell Us? 2
An Outline of the Immunocytochemistry Procedure 4
What Is Included in This Book? 5

2 Antibodies 7
Introduction 7
Antibody Molecules 8
Making Antibodies 10
Talking About Antibodies 13
Finding and Getting Antibodies 14
Choice of Primary (1°) Antibodies 15
Antibodies Handling and Storing 16
Recommended Storage Freezer, –20°C 16
Recommended Storage Refrigerator, 4°C 16

3 Sample Preparation/Fixation 17
Introduction 17
Fixation Theory 18
Chemical Fixatives 19
Vehicle 22
Applying Fixatives 24
Dissecting the Area of Interest 25
Protocol – Fixation 26
Components for Paraformaldehyde Fixative 26
Procedure 27
Perfusion Procedure 27
Perfusion Equipment 28
Drop-in-Fixation 28

4 Tissue Sectioning 29
Introduction 29
Embedding Tissue by Freezing 30

Theory of Freezing Tissue   30
Freezing Tissue   32
Cryostat Sectioning   33
Tissue Processing   37
Vibratome, Freezing Microtome, and Microwave   39
Fresh Frozen Tissue   41
Embedding Tissue with Paraffin   41
Cryostat Protocol   42

5   **Blocking and Permeability**   45
Introduction   45
Nonspecific Antibody Binding to Tissue and Cells   45
Blocking for Nonspecific Antibody Binding   47
Permeabilize Tissue and Cells to Allow Antibody Penetration   49
Effects of Blocking Agents on Antibody Penetration   51
Combined Incubation Step   53

6   **Labels for Antibodies**   55
Introduction   55
Fluorescence Theory   56
Four Generations of Fluorescent Labels   58
Immunocytochemistry Fluorophores and Flow Cytometry   59
Choosing Fluorochromes   61
Enzyme Theory   61
Enzyme Substrates   61
Particulate Label   63
Choice of Fluorescent or Enzymes for Immunocytochemistry   64

7   **Application Methods**   65
Introduction   66
Direct Immunocytochemistry   66
Direct Immunocytochemistry Advantages   67
Direct Immunocytochemistry Disadvantages   67
Indirect Immunocytochemistry   67
Indirect Immunocytochemistry Advantages   68
Indirect Immunocytochemistry Disadvantages   68
Avidin–Biotin Molecules   68
Direct Avidin–Biotin Immunocytochemistry   69
Direct Avidin–Biotin Method Advantages   70
Direct Avidin–Biotin Method Disadvantages   70
Indirect Avidin Biotin Immunocytochemistry   70
Indirect Avidin–Biotin Advantages   71
Indirect Avidin–Biotin Disadvantages   71
Avidin–Biotin Complex (ABC) Immunocytochemistry   71
Avidin–Biotin Complex (ABC) Advantages   73
Avidin–Biotin Complex (ABC) Disadvantages   73

Tyramide Signal Amplification (TSA) Immunocytochemistry   73
  Tyramide Signal Amplification Advantages   74
  Tyramide Signal Amplification Disadvantages   75
  ABC with TSA   75
    ABC with TSA Advantages   77
    ABC with TSA Disadvantages   77

8 Controls   79
  Introduction   79
  Three Immunocytochemistry Controls   79
    1. 1° Antibody Controls   80
    2. 2° Antibody Controls   84
    3. Labeling Controls   85

9 Method and Label Decision   89
  Introduction   89
  Choose Application Label and Method   89
  Experimental Design Chart   93

10 Single Antibody Procedure   97
  Introduction   97
  Experimental Design Chart   98
  Incubation Conditions   98
  Antibody Dilutions 100
  Antibody Dilution Matrix 102
  2° Antibody Controls 102
  Rinses 104
  Mounting Media 105
  Final Procedure 106
    Steps in a Single 1° Antibody Indirect
    Immunocytochemistry Experiment 106
    Steps in a Single 1° Antibody Immunocytochemistry
    Experiment for Ag A 107

11 Multiple Antibodies Different Species 111
  Introduction 111
  Combining Two 1° Antibody Incubations 112
  Experimental Design Chart 112
  Designing 2° Antibody Controls 113
  Rules for Multiple Label Experiments 113
  Complete Final Procedure 115
    (D) Block and Permeabilize 116
    (E) Rinse after Block and Permeabilize 116
    (F) 1° Antibodies 117
    (G) Rinse After 1° Antibody 117
    (H) 2° Antibody 117
    (I) Rinse After 2° Antibody 117

12 Multiple Antibodies from the Same Species 119

Introduction 120

Combine Two 1° Antibodies from the Same Species
with Block-Between Method 120

Experimental Design Chart for Block-Between Method 122

Design the 2° Antibody Control for the Same Species
with Block-Between 124

Final Procedure for Two 1° Antibody Same Species
with Block-Between 127

   (A) Prepare Cell Culture 127

   (B) Fix Culture 127

   (C) Block and Permeabilize 127

   (D) Rinse After Block and Permeabilize 128

   (E) Incubate First 1° Antibody 128

   (F) Rinse After First 1° Antibody 128

   (G) Incubate First 2° Antibody 128

   (H) Rinse After First 2° Antibody 128

   (I) Block Antibodies in First Set 128

   (J) Incubate Second 1° Antibody 128

   (K) Rinse After Second 1° Antibody 129

   (L) Incubate Second 2° Antibody 129

   (M) Rinse After Second 2° Antibody 129

   (N) Mount Coverslip 129

   (O) Examine in Microscope 129

   (P) Evaluate Results 129

Combine Two 1° Antibodies from the Same
Species with Zenon 130

Experimental Design Chart for the Same Species with Zenon 130

Design the Antibody Control for the Same Species with Zenon 133

Final Procedure for Two 1° Antibody from the Same Species
with Zenon 135

   (A) Prepare Cell Culture 135

   (B) Fix Culture 135

   (C) Block and Permeabilize 136

   (D) Rinse after Block and Permeabilize 136

   (E) Prepare the Zenon Reagents 136

   (F) Incubate with Labeled Antibody(ies) 136

   (G) Rinse After Antibody Incubation 136

   (H) Fix with 4% Paraformaldehyde 136

   (I) Rinse after Antibody Incubation 137

   (J) Mount Coverslip 137

   (K) Examine in Microscope 137

   (L) Evaluate Results 137

13 Fluorescent Microscopy and Imaging 139
    Introduction 139
    Filter Sets in Fluorescence Microscopy 140
    Fluorescent Bleed-Through 142
    Fluorescence Quench and Photobleach 145
    Image Parameters – Contrast and Pixel Saturation 146
    Ethics of Image Manipulation 148
        Do 149
        Do Not 149

14 Troubleshooting 151
    Introduction 151
    Procedural Errors 152
    Method of Troubleshooting 152
        Case No. 1 153
        Case No. 2 156
        Case No. 3 158
        Case No. 4 164
        Case No. 5 167
    Troubleshooting Unique to Multiple Primary Antibodies 173
    Bad Antibodies 173
        Bad 1° Antibodies 173
        Bad 2° Antibodies 174

15 Electron Microscopic Immunocytochemistry 175
    Protocol – Pre-embedding Electron Microscopic Immunocytochemistry 175
    Introduction 175
    Need for Electron Microscopic Immunocytochemistry 176
    Pre-embedding Electron Microscopic Immunocytochemistry 178
    Postembedding Electron Microscopic Immunocytochemistry 181
    Choice of a Method 185
        Advantages and Disadvantages 185
    Protocol – Pre-embedding Electron Microscopic Immunocytochemistry 185
    Solutions 186
        Stock Solutions to Make Ahead and Store 186
        Solutions Made on the First Day of the Experiment 187
        NPG Silver Enhancement Solution and Silver Lactate 188
        Test Strip 189

Appendix 191

References 199

Glossary 203

Index 213

About the Author

Richard W. Burry, PhD, is the Director of the Campus Microscopy and Imaging Facility (CMIF) at The Ohio State University and an Associate Professor in the Department of Neuroscience at Osu's College of Medicine. He received a BS from Beloit College in Beloit, Wisconsin and a PhD from the University of Colorado Medical Center, Denver, Colorado. He has received NIH grants, NSF grants, and industry contracts leading to over 50 publications and numerous presentations at scientific meetings. He has been Secretary and President of the Histochemical Society and organized the 6th Joint Meeting of the Japan Society for Histochemistry and Cytochemistry and the Histochemical Society, in 2002. Dick has also been an Associate Editor for the Journal of Histochemistry and Cytochemistry since 1999. In 2009, at the 60th Annual Meeting of the Histochemical Society in New Orleans, Dick received the Carpenter–Rash Award for outstanding contributions and service to the Histochemical Society.

Chapter 1
Introduction

Keywords Immunohistochemistry · Antibody labeling · Fluorescence micros-copy · Fluorescent immunocytochemistry · Fluorescent immunohistochem-istry · Indirect immunocytochemistry · Immunostaining

Contents

What Is Immunocytochemistry? 1
What Can Immunocytochemistry Tell Us? 2
An Outline of the Immunocytochemistry Procedure 4
What Is Included in This Book? 5

What Is Immunocytochemistry?

Immunocytochemistry is the use of antibodies for identifying proteins and molecules in cells and tissues viewed under a microscope. Immunocytochemistry harnesses the power of antibodies to give highly specific binding to unique sequences of amino acids in proteins. Perhaps the most exciting part of using antibodies is that new antibodies can be generated on an as-needed basis, thus providing a constant source of new reagents. Scientists are constantly generating new antibodies to specific parts of molecules thus driving continual evolution of immunocytochemistry. Identifying the location of antibodies in cells is based on availability of labels that is, itself, rapidly advancing. As time passes, immunocyto-chemistry continues to respond to new development of labels and advanced methods of labeling molecules.

If the terms immunocytochemistry and immunohistochemistry seem similar then here is why. Many years ago, immunocytochemistry was defined as the use of antibodies to study cells in the form of cultures or smears from animals. Immunohistochemistry, on the other hand, was defined as the use of antibodies to study paraffin sections from human tissue. Today, immunohistochemistry is still

R.W. Burry, *Immunocytochemistry*, DOI 10.1007/978-1-4419-1304-3_1, © Springer Science+Business Media, LLC 2010 the use of antibodies in paraffin sections in human pathology, but the definition of immunocytochemistry has changed. *Immunocytochemistry is the use of antibodies in animal research with cells and tissues fixed in paraformaldehyde.*

This new definition of immunocytochemistry derives from advances in antibody-labeling methods in recent years. These advances resulted from specific needs in animal research. Initially, formalin-fixed paraffin sections were used for immuno-histochemistry; however, results were inconsistent. In most cases, the antibody did not label anything or it labeled too many cells and was dubbed "over fixed." This problem led to the development of the *epitope retrieval or antigen retrieval* methods, where sections of tissue are treated with heat in buffers before antibody incubations. Unfortunately, epitope retrieval methods can be unique from antibody to antibody and also, for the same antibody, from tissue to tissue. Epitope retrieval is complicated and best avoided. For animal research, a simple method was then developed where tissue was fixed in paraformaldehyde and not formalin or alcohol and subsequently frozen sections were cut on a cryostat. This eliminated the steps of dehydration, embedding in paraffin, rehydration after sectioning, and epitope retrieval before antibody incubation. This was a major breakthrough.

Today, for research with animal tissue and cell cultures, the standard has become fixation in paraformaldehyde, with animal tissue sectioned in a cryostat, and then incubation of sections and cultures with antibodies. This book focuses on introducing the methods of immunocytochemistry for biomedical scientists. These chapters may be read in order for a complete understanding of immunocytochemistry, or the chapters may be read individually for information about specific topics. The book is designed to help the novice perform experiments, solve problems, get results, and understand more advanced texts when more advice is needed.

What Can Immunocytochemistry Tell Us?

Immunocytochemistry harnesses antibodies that are specific reagents and which allow unique detection of proteins and molecules. Using antibodies requires specific methods, labels, and controls. Performing immunocytochemistry experiments requires some basic knowledge of biology.

Much of the data collected in biomedical research today results from biochemical and molecular methods, where many cells are pooled for analysis. For example, enzyme assay of the liver will give values that, when repeated, should be statistical similar and should provide reliable average values with standard errors. When this and similar methods pool many cells for analysis, they are broadly defined as *"population studies"* (Fig. 1.1a). However, problems result, because not all liver cells might have the specific enzyme of interest. So changes found with the enzyme assay might be due to enzyme activity in all of the liver cells or might be due to enzyme activity in only some of the cells. Rather than assuming all of the cells in the liver have the enzyme, the complementary approach is to look at the cells with morphological methods.

Morphological approaches in biomedical research can include a wide range of microscopes, but today typically employ immunocytochemistry that can give us information about individual liver cells containing the specific enzyme. Immunocytochemistry uses antibodies to bind proteins and labels to show protein's location. If, for example, the enzyme is a marker for inflammation, then the location of cells with this enzyme tells us which cell types have the inflammatory response. Thus, immunocytochemistry methods are broadly defined as *"individual studies"* of single cells or cell groups. The resulting data tell us about location of the enzyme.

To look at how immunocytochemistry (an individual study) has advantages over an enzyme assay (a population study), let us compare the types of results from these two approaches. To determine the enzyme level, the liver is ground up and a specific biochemical enzyme assay is performed (Fig. 1.1a). At different time points, the level of enzyme activity increases significantly as seen by the small errors shown on the graph (Fig. 1.1a). However, it is easy to assume that all cells in the liver have the enzyme (Fig. 1.1b) detected in the biochemical assay. In reality, however, the

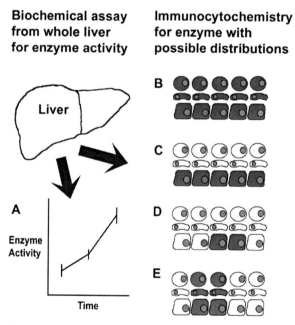

Biochemical assay from whole liver for enzyme activity

Immunocytochemistry for enzyme with possible distributions

Fig. 1.1 Morphological and biochemical studies: (**a**) One biochemical approach to study enzymes is to analyze the activity levels with results plotted on a graph and to include error bars from multiple assays. The morphological approach gives information about where the enzyme is located. Three different types of liver cells are shown here as circular, elongated, and rectangular. (**b**) The enzyme (dark cells) can be located in all the different types of liver cells. (**c**) More likely the enzyme is found in only one cell type, the rectangular cells. (**d**) As a result of disease, the enzyme may be expressed in only a small number of cells in a single cell type. (**e**) Following an injury, the enzyme may be expressed in multiple cell types located near the injury sites

liver is composed of several different cell types, each with different functions and consequently some cells may not contain the enzyme.

To explore the possibility that only a subpopulation of liver cells has the enzyme immunocytochemistry, an individual study is performed. Results could show one of several patterns of distribution for the enzyme. The enzyme could be found in one cell type in the liver (Fig. 1.1c). But more realistic scenario is that the enzyme is found only in few cells of a specific cell type due to local injury (Fig. 1.1d). If injury is causing the enzyme activity, then most likely that expression of the enzyme will be seen in several cell types near the injury site (Fig. 1.1e). Thus, immunocytochemistry gives us valuable information about the location and number of cells expressing the enzyme. *The important point here is that biochemical and immunocytochemistry data are complementary; neither can replace the other.*

Another example of a population study that uses antibodies is flow cytometry. Isolated cells must be dissociated from tissues or cultures and labeled with fluorescent antibodies specific for a subpopulation of the cells. In flow cytometry, cells pass rapidly past a detector that measures the amount of fluorescence for each cell. The size of cells and the amount of fluorescence can be plotted and analyzed. Even though this method makes use of antibodies, it is a population study because it determines the number of isolated cells bound to an antibody. Flow cytometry identifies different populations of isolated cells, but it cannot show the location of these labeled cell in tissues, which can be done only with immunocytochemistry.

An Outline of the Immunocytochemistry Procedure

Here then is how it all works together. Immunocytochemistry takes tissue sections or culture cells and incubates them with antibodies. The experimental needs determine the exact order of antibodies incubations and the specific labeling of the antibodies. The general steps in a single primary ($1°$) antibody indirect immunocytochemistry experiment include the following:

(A) Prepare samples
(B) Fix tissue or cells
(C) Embed, section, and mount tissue
(D) Block and permeabilize
(E) Rinse after block and permeabilize
(F) Incubate with $1°$ antibody
(G) Rinse after $1°$ antibody
(H) Incubate with $2°$ antibody
 (I) Rinse after $2°$ antibody
(J) Mount coverslip
(K) Examine in microscope
(L) Evaluate results

Each chapter follows the order of these procedure steps and explains different options. The goal is to give enough information to design a procedure for a particular experimental need. In Chapter 9, Decision Method and Label, an Experimental Design Chart is presented that guides the decisions on reagents for specific experiments.

What Is Included in This Book?

It would be great if a single procedure was universal and could meet the needs of each individual scientist! However, what typically happens is that each scientist has unique needs. Scientists looking for apoptotic enzymes in cardiac muscle, neuropeptides in the brainstem, and RNA-binding proteins in cultured cells cannot all use a single procedure.

But when scientists understand the principles and methods of immunocytochemistry, they can and do design experiments with few problems. In addition, they are able to solve the problems they might encounter. This book is designed to provide the necessary concepts for understanding and practicing successful immunocytochemistry. Methods and principles described in this book are given in sufficient detail for an essential understanding. Methods that are of historical interest only are not included in detail. For example, there is no discussion in this book of the peroxidase anti-peroxidase (PAP) method of Ludwig Sternberger (Sternberger et al., 1970) that initially revolutionized increased sensitivity for immunocytochemistry. PAP is an important method, but it is not used today because other methods are preferred. Instead this book guides the novice user on currently popular, productive methods of immunocytochemistry. For a historical approach to immunocytochemistry including advanced methods not covered here, several excellent books are available (Larsson, 1988; Polak and Van Noorden, 2003; Renshaw, 2007).

This book is intended for scientists who are working on research animals and cultured cells. The procedures described here give the best results with the easiest methods. Note that many older procedures and reagents are still used today, but they give less than ideal results. For example, the fluorophore FITC was the first fluorophore used for immunocytochemistry by Albert Coons in 1942, when he invented this field. Since then, three new generations of fluorescent compounds (Chapter 6, Labels) with improved photobleaching properties have evolved making FITC of historical interest for immunocytochemistry.

This book is organized like the planning of an immunocytochemistry experiment. The initial chapters explain the choice of reagents and methods in different processing steps such as fixation and sectioning. The later chapters support the design of a specific experiment. There are charts and lists for decision making. The last chapters deal with microscopy and image collection. For truly novice users of immunocytochemistry, plan a day or so for reading and planning before taking this book to the laboratory.

For more experienced users, individual chapters can be used to guide a specific part of the immunocytochemistry method. For example, if the user needs to understand the choices of detergents used to open the cells and allow for penetration of antibodies, then start with Chapter 5, Block and Permeability. The Experimental Design Chart is introduced in Chapter 9 (Methods and Label Decision) and helps organize the choosing and testing of reagents needed for each protocol. Chapter 9 includes some examples of completed charts to provide an idea as to their function. Protocols are listed at the ends of appropriate chapters on the corresponding topic.

Chapter 2
Antibodies

Keywords Immunohistochemistry · Antibody labeling · Fluorescence micro-scopy · Fluorescent immunocytochemistry · Fluorescent immunohistochem-istry · Indirect immunocytochemistry · Immunostaining

Contents

Introduction 7
Antibody Molecules 8
Making Antibodies 10
Talking About Antibodies 13
Finding and Getting Antibodies 14
Choice of Primary (1°) Antibodies 15
Antibodies Handling and Storing 16
  Recommended Storage Freezer, –20°C 16
  Recommended Storage Refrigerator, 4°C 16

Introduction

An antibody (Ab) is the key reagent of immunocytochemistry. To use antibodies effectively, consider their structure, function, and generation. Such basic knowledge about antibodies is essential to succeed in identifying suitable experimental design, finding antibodies, and trouble-shooting problems.

Immunocytochemistry takes advantage of three properties of antibodies:

1. Antibodies uniquely bind to a protein or other molecule.
2. Antibody binding to molecules is essentially permanent at physiological conditions.
3. New antibodies can be made tailored to new interesting molecules.

Antibody Molecules

An immune response generates antibodies or proteins called *immunoglobulins (Igs)*. Antibodies are further classified into multiple *isotypes or classes* (Table 2.1). In immunocytochemistry, the *IgG* isotype is preferred because its generation and binding is more consistent. IgM antibodies can be used if no other isotype is available. The IgG molecules can be broken down into four *subclasses*, IgG1, IgG2, IgG3, and IgG4. In immunocytochemistry experiments, these subclasses do not matter for most species of animals, but they are important for antibodies generated in mouse monoclonal antibodies (IgG1, IgG2a, IgG2b, and IgG3), as we will see in later chapters.

Table 2.1 Ig isotypes

Antibody isotype	Action
IgA	Gut, respiratory, urinary response
IgD	Initial immune system response
IgE	Response to allergens – histamine release
IgG	Immune response to invading pathogens
IgM	Early immune response for pathogens

In using antibodies, knowledge of the IgG structure is important (Fig. 2.1). *IgG has a constant region and a variable region*. The *constant region* contains species-specific sequences and the *Fc portion* that binds an Fc receptor (Fig. 2.1, clear end), which is found on circulating white cells, macrophages, and natural killer cells. The Fc portion also has species-specific sites that are unique to the animal species

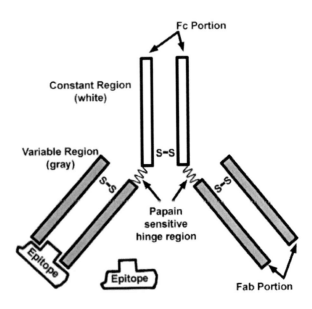

Fig. 2.1 The antibody. An IgG antibody has a single constant region (*white*) with the Fc portion and the species-specific antigens. The variable region (*gray*) contains the Fab portion that binds the epitope portion of the antigen. The small protein, only in the variable region, is known as the light chain; the large protein that is part of the constant and variable region is the heavy chain. The IgG can be digested by the protease enzyme, papain, into an Fc end (constant end) and a Fab end (variable end)

in which the antibody was generated. Thus, generation of an antibody against IgG from rabbit will result in antibodies that bind the constant region from rabbit IgG only and not, for example, from mouse IgG.

Immunocytochemistry uses antibodies against IgGs. Antibodies or IgG molecules are generated to other IgG molecules by injecting purified IgG molecules from one species into another species. In the case of mouse IgG injected into rabbit, it will produce rabbit anti-mouse IgG antibodies. Antibodies made against an IgG will only bind to the constant region or Fab region of the IgG.

The variable end of the antibody contains the unique epitope-binding regions that give each antibody its specificity (Fig. 2.1, gray end). This *variable region* is the fraction antigen binding *(Fab)* portion. The unique configuration of the Fab specifically binds the epitope. When an antigen is injected into a rabbit, the resulting antibodies against the antigen have Fab portions that are unique to the antigen, but the rest of the IgG is similar to other IgG molecules.

Each IgG antibody has two Fab ends, which can bind to two identical epitopes at the same time. The advantage of this *bivalent epitope binding* is that it can amplify the epitope detection. The orientation of the two epitopes is not restricted as there are hinge regions (Fig. 2.1) in the IgG molecule that connect the Fab portion to the Fc portion of the IgG. The hinge region allows movement and rotation of each individual Fab, thus facilitating binding to adjacent identical epitopes.

Heavy chains or long protein (Fig. 2.1, light and dark bars connected by a papain-sensitive hinge) and light chains or short protein (Fig. 2.1, short dark bar) IgG molecules are made of two proteins that are held together by disulfide bonds of the amino acid cysteine (Fig. 2.1; S–S between bars).

The enzyme, papain, can digest the hinge regions of IgG and can generate two identical Fab portions and one Fc portion. The individual Fab portion can be used for immunocytochemistry, where single epitope-binding region is needed without species-specific binding.

An *antigen* is a protein, peptide, or molecule used to cause an immune response in an animal. The animal responds by making antibodies to individual *epitopes located on the antigen.* An individual *antigen has multiple epitopes* that can generate antibodies. In Fig. 2.2, the "&" represents an antigen and the light gray areas on the edge represent individual epitopes. *An epitope can be an amino acid sequence on a*

Antibody clone numbers   1   2   3   4   5   6
Antibodies to many
epitopes on antigen "&"

Fig. 2.2 Antibody generation. Antigens are the molecules injected into animals that generate antibodies ("&" is an antigen). Epitopes are small parts of antigens that generate a specific antibody (short *gray lines* on "&" are epitopes). Here, six antibodies (small Ys) are generated to epitopes on the antigen "&." Each different antibody is from a clone of B-cells (with numbers); each B-cell produces antibodies to only one epitope; some clones can produce antibodies to the same epitope as other clones (clones No. 1 and No. 4)

denatured peptide or a several sequences on the surface of a folded protein. Animals frequently generate multiple antibodies to the same epitope (Fig. 2.2, clones 1 and 4). Also, an epitope on one protein might also exist on a different, unrelated protein because it has the same sequence or the same surface configuration.

Making Antibodies

An animal injected with an antigen will generate multiple antibodies to many epitopes. Antibodies are produced by B-cells and *a single clone of B-cells produces antibodies to only a single epitope.* Once a B-cell begins producing a single type of antibody, it will divide and give rise to many B-cells, all producing that single antibody to just one epitope; this is called a *B-cell clone.* Sometimes there are multiple clones of B-cells that produce antibodies to a single epitope (Fig. 2.3, clones 1 and 4). Parts of injected proteins and molecules make better antigens than others. As a result, some proteins do not generate many antibodies. An example is G-coupled receptors, a class of membrane receptors, that do not generate antibodies well.

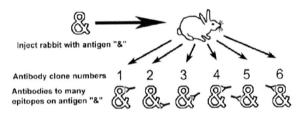

Fig. 2.3 Polyclonal antibodies. An animal injected with an antigen will generate B-cell clones that can produce antibodies to multiple epitopes. The serum from the animal has different antibodies to these multiple clones, thus the name, polyclonal

Polyclonal antibodies contain multiple clones of antibodies produced to different epitopes on the antigen. In Fig. 2.3, the serum from an immunized rabbit contains antibodies from six clones of B-cells. In serum from the rabbit, the six different clones of antibodies will increase the labeling of the antigen because there are multiple epitopes on the antigen. Polyclonal antibodies are in the form of serum from animals and are made in different species of large animals (rabbit, donkey, goat, sheep, and chicken). Chicken polyclonal antibodies are purified from unfertilized egg yolks, with the advantage that eggs are easy to collect and large amounts of an antibody can be isolated from a single chick.

Advantages of Polyclonal Antibody

- Multiple clones give high levels of labeling for a single antigen because they contain many antibodies to different epitopes on the same protein.

Disadvantages of Polyclonal Antibody

- Shared epitopes on different proteins can label multiple proteins that are not the antigen protein.
- Obtaining the antibody depends on a living animal and the ultimate death of the rabbit means no more antibody.
- When a new rabbit is immunized with the same antigen, the exact epitopes generating antibodies will be different and a different number of clones are generated.

Monoclonal antibodies, originally from one mouse, contain a single antibody from one clone of B-cells to a single epitope on the antigen. This procedure was first described by Georges Kohler and Cesar Milstein, for which they received the Nobel Prize in 1984. Monoclonal antibodies are made by immunizing a mouse, and when antibodies are produced, the spleen of immunized mouse is removed (Fig. 2.4). The spleen cells are dissociated including the B-cells producing antibodies (Fig. 2.4, different gray levels). Because B-cells will not divide in culture, they must be fused with a continuously dividing cell line that produces antibodies. Such a cell line is the mouse myeloma cell line.

The spleen cells are fused with mouse myeloma cells to become a continuous hybridoma cell line. A continuous hybridoma cell line with multiple B-cell clones produces many different antibody clones indicated by the different gray levels of

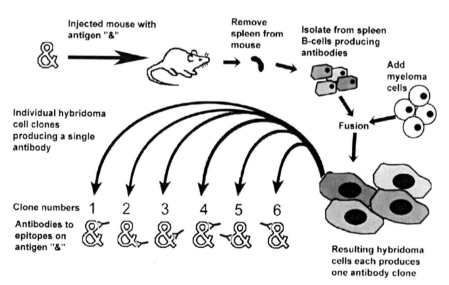

Fig. 2.4 Monoclonal antibodies. After injecting the antigen and generating several clones of antibodies, the spleen containing B-cells is removed. Hybridoma cells are made by fusing spleen B cells with a myeloma cell culture line. To isolate the individual hybridoma cells producing one clone of antibody, the mixed hybridoma culture is highly diluted and plated in 96-well plates with one cell or less per well

the cells in Fig. 2.4. Next, the population of *hybridoma cells producing many antibodies is cloned* in 96-well plates and each single B-cell clone of cultured cells produces one antibody. Individual clones producing a separate antibody are named by location in the 96-well plate (e.g., 5B12 plate 5, row B, column 12). One mouse spleen can give many different antibodies to different epitopes on the same antigen. Monoclonal antibodies are raised in either tissue culture media, called *supernatant*, or generated from hybridoma cells injected into the peritoneal cavity (abdominal cavity), called *ascites fluid*. Until recently, all monoclonal antibodies were generated exclusively from mice because of the limitations with generating good myeloma cell lines for other species of animals. Rabbit monoclonal antibodies are now available because a good rabbit myeloma cell line is now available. Rabbit monoclonal antibodies have high sensitivity and excellent response to antigens from mouse tissue.

As a result of the popularity of rabbit monoclonal antibodies, confusion exists when using the term monoclonal. Previously, monoclonal antibodies were always from mouse and so detection systems were always based on binding to mouse monoclonal antibodies. Now with the popularity of rabbit monoclonal antibodies, it is not possible to use the term monoclonal to identify the species of the antibody.

Advantages of Monoclonal Antibodies

- Single clone monoclonal antibodies bind to a single epitope, which is selected for high specificity for the antigen.
- Different clones of antibodies can be generated to different epitopes on a single antigen.
- Single clone can be generated to a posttranscriptionally altered protein (e.g., phosphorylated amino acid).
- Clones to an epitope shared with multiple proteins (gene products) can be rejected.
- The same antibody can be generated indefinitely from cultured hybridoma cells in a process that creates a stable reagent.
- The identical clone sold by different companies will be the same antibody.

Disadvantages of Monoclonal Antibodies

- Much work is required to generate a successful monoclonal antibody, especially in the cloning and selection process.
- Low levels of labeling occur because the monoclonal antibody binds an infrequent epitope on a protein or binds with low affinity.
- Monoclonal antibodies are mostly from mice because of a strong myeloma cell line.

Talking About Antibodies

Terminology is important in describing the source and specificity of antibodies used in immunocytochemistry. *The species used to generate antibodies are used to differentiate antibodies.* An antibody generated in rabbit to the protein tubulin would be a "rabbit anti-tubulin antibody." With both mouse and rabbit being used to make monoclonal antibodies, the species of the animal generating the monoclonal antibody must be stated, and not simply "monoclonal" to mean antibodies produced in mouse. To identify an antibody, use the species of animal where the antibody was generated and not the term monoclonal.

Concentrations of IgG in

serum is 1–10 mg/ml;
ascites is 1–2 mg/ml; and
supernatant is 0.4–1 mg/ml

Antibodies can come in a variety of forms and purities. Polyclonal antibodies can come as *whole serum* or as purified antibodies with an IgG concentration of 1 mg/ml. Monoclonal antibodies come as isolated tissue culture media from hybridoma cells called *supernatant*. The antibody from supernatants is between 50 and 100 μg/ml, which means that the working antibody dilution for immunocytochemistry will be lower than whole serum. In addition, monoclonal antibodies can be *ascites fluid* giving antibodies that are highly concentrated of 1 mg/ml. Today, generation of ascites may be restricted by federal regulations for care of research animals.

To increase the purity or to concentrate an antibody solution, it may be purified. Purification is done with a range of techniques applied to whole serum, supernatant, or ascites fluid. At the first level, the *purified Ig* will be separated from other serum proteins and will select all IgGs including the IgG of interest and other IgG molecules. These purification steps can be done by using ammonium sulfate to precipitate the Ig molecules or it can be done by binding antibodies to a Protein A and/or Protein G columns. Proteins A and G are produced by the bacteria, *Staphylococcus aureus,* and bind to different species and subclasses of antibodies by the Fc receptor. After the antibodies have attached, they are washed out by changing the buffer.

The next level of purification is *affinity purification*, where the antigen is available and can be bound to a column, the serum or supernatant is passed over the column binding to the antigen. The antibodies are washed off with low salt and detergent-containing buffers. The third level of purification is used if the antigen is not available. A band from a gel containing the protein of interest can be cut out and used to purify the antibody. Affinity-purified antibodies are, in theory, the best because they have bound to the antigen. However, some of the strongest binding antibodies cannot be eluted from the affinity columns and recovered, so there is controversy about the value of affinity purification.

All antibody solutions should be clear and free of particles or other precipitated material essential to eliminating background labeling in immunocytochemistry. IgG purification removes any particulate material from the whole serum, supernatant, or ascites fluid that could cause background.

Finding and Getting Antibodies

Selecting an antibody can be a daunting task. Most commonly, antibodies will be purchased from a vendor. There are hundreds of vendors selling antibodies. Finding good antibodies is best done by looking in journal articles or by getting a recommendation from someone who is using a specific antibody. Regardless, to successfully use an antibody requires information about that antibody. What follows is a list of items that should be available from vendor in the *product information* for all antibodies.

Catalogue information – The catalogue number and the price.

Description or background – The name of the antigen, its molecular weight, alternative names, and something about antigen's function.

Antibody type or host – The name of the species used to generate the antibody, the isotype of the antibody, and the clone name/number, if the antibody is monoclonal. If the antibody is a Fab fragment, it should be stated.

Source of antigen – The nature of the injected antigen (protein, peptide from a specific sequence) and the species of the antigen. Sometimes antibodies to specific parts of molecules are needed (e.g., the extracellular domain, a specific sequence of amino acids or a posttranslational modification).

Packaging, product, or purification – The amount of the liquid product, the concentration of antibody in the product (1 mg/ml is ideal), additives (e.g., sodium azide, glycerol), the source (e.g., whole serum, supernatant, ascites), and purification, if any.

Specificity – A description of how the producer determined that the antibody binds only the listed antigen. Most of the time this is a western blot (with a blot shown), but it can be immunocytochemistry (with an image shown). Sometimes data are included about binding to other related proteins or to posttranslationally modified (e.g., phosphorylated) proteins. Some vendors who use a peptide for making an antibody will also sell the peptide for an absorption control. More information about specificity is discussed in the Chapter 9, Controls. Frequently, there are no data given for the specificity.

Uses or application – The methods where the antibody has been tried are generally any of the following: immunohistochemistry (IHC), immunocytochemistry (ICC), or immunofluorescence (IF); western blot (WB) or immunoblot (IB); and immunoprecipitation (IP). This information should also include recommended dilutions for the listed applications. When purchasing of antibodies for immunocytochemistry, focus only on those tested with immunocytochemistry.

Species reactivity – The species where the antibody binds the antigen. If the antibody does not react to the species of your tissue or has not been tested against the species you are using, do not use it. Some vendors will send a sample to test for a new species. Alternatively, the species that the antibody does not react with might also be listed. This is important because not all antibodies will bind equally to antigens from different species.

Protocols – The recommended protocol that can be useful to understand any unique fixation, detergent, blocking, or incubation conditions. Many vendors will list references for papers that use their antibody rather than list a protocol.

Some antibodies generated in research labs are available from individuals. Antibodies obtained from noncommercial sources will not have the level of documentation expected from commercial sources; however, sometimes it is available. The advantage of antibodies from individual scientists is that if the antibody does not work, the person giving you the antibody will usually help you to get the antibody to work. A disadvantage is that the antibody obtained once may not be available when you need more to complete a study.

Choice of Primary (1°) Antibodies

In immunocytochemistry, the antibody that binds to the antigen is called the *primary antibody (1° antibody)*. In searching for 1° antibodies, the first source of information is the published literature. The best search will show several antibodies to the antigen presenting a choice of antibody possibilities. If there are several antibodies available, the factors that determine your choice should be, in the following order: recommended in literature, product literature recommendation for immunocytochemistry, high specificity for the antigen of interest including same species as your tissue, species where the antibody was generated (important when used for multiple antibody experiments), and price.

Note that publication of research using results with an antibody requires that the author includes in the manuscript the source of the antibody and how its specificity was determined. List the company, the product name, product number, lot number, the supplier-specificity results, and any specificity experiments that were performed. Antibodies from individuals generally require negotiation with the individual because the individual might require that he/she be designated a co-author on any paper using his/her antibody. Alternatively, be sure that you acknowledge the individual as the source of the antibody in any manuscript.

Antibodies Handling and Storing

Antibodies are variously supplied in specific forms and shipped frozen, on ice or at room temperature depending on the antibody. Antibodies should be stored as indicated by the vendor or supplier. Antibodies are reasonably stable but can be damaged by repeated freezing and thawing, extreme pH, and high salt environments.

Recommended Storage Freezer, –20°C

- Repeated freezing and thawing will denature antibodies. Damage is reduced by diluting in 30% glycerol.
- Aliquot antibodies so they will be thawed once.
- Passive or nonfrost-free freezers at –20°C or freezers at –70°C. Not recommended are frost-free freezers that have circulating air which will dehydrate frozen antibody in months.

Recommended Storage Refrigerator, 4°C

- Prevents damage due to freeze–thaw.
- Nonfrost-free refrigerators. Frost-free refrigerators will dehydrate and concentrate and eventually dry the antibody solutions.
- Add 0.02% sodium azide to inhibit growth of bacteria (many companies do this). Note: sodium azide will inhibit and enzyme used for immunocytochemistry, horseradish peroxidase (HRP).

Storage of antibodies is controversial. Hint: After years of antibody use, we recommend storing antibodies in small aliquots in microfuge tubes in a –70°C freezer. These aliquots of about 10 μl contain enough liquid for one or two uses. These antibodies will last for many years at this temperature and will not dry because there is no circulating cold air. The downside is that this process requires a lot of space and a great record system to keep track of where individual antibodies are located.

Chapter 3
Sample Preparation/Fixation

Keywords Immunohistochemistry · Antibody labeling · Fluorescence micro-scopy · Fluorescent immunocytochemistry · Fluorescent immunohistochemis-try · Indirect immunocytochemistry · Immunostaining

Contents

Introduction 17
Fixation Theory 18
Chemical Fixatives 19
Vehicle 22
Applying Fixatives 24
   Dissecting the Area of Interest 25
Protocol – Fixation 26
   Components for Paraformaldehyde Fixative 26
Procedure 27
   Perfusion Procedure 27
   Perfusion Equipment 28
   Drop-in-Fixation 28

Introduction

Sample preparation is one of the most important steps in immunocytochemistry because it generally receives the least amount of planning. Unfortunately, few researchers understand how critical the first few steps in an immunocytochemistry experiment are. In fact, the quality of the cells and tissue and the ability to get good results are *totally* dependent on initial fixation. To put sample preparation in perspective, perhaps the best thing to do is focus on the conclusion of the experiment – the quality of final microscopic image. Good images only come from cells and tissues that are fixed properly.

R.W. Burry, *Immunocytochemistry*, DOI 10.1007/978-1-4419-1304-3_3,
© Springer Science+Business Media, LLC 2010

All tissue and cultures for immunocytochemistry must be fixed to preserve them.
Unfixed cells and tissue degenerate quickly, leaving nothing to be seen. In fact, once
the tissue culture medium or blood supply is removed, the process of degeneration
begins. Therefore, it is crucial that immediately upon removing the tissue from the
animal or the cells from the culture medium begin the preservation process with a
fixative.

Fixation Theory

Fixation is the stabilization or preservation of cells and tissue as close to life-like as
possible. All fixation procedures change the tissue they are preserving, but the key
is to find the least amount of change for immunocytochemistry.

Ideally for immunocytochemistry, it would be nice to examine live cells and
tissues without any fixation. The problem with looking at cells and tissue in the
microscope is that the sample must be *thin enough to examine in the microscope
and it must be stable enough so as not to deteriorate while being examined.* These
criteria require the use of fixatives to preserve the cells and tissues.

In theory, fixation for microscopy is based on the need to obtain images of tissues
and cells as they were when living, with no changes or distortions. However, fixation
has the potential of introducing changes in cells and tissues. The purpose of this
chapter is to guide sample preparation to minimize unwanted changes to cells and
tissues.

The following are the *criteria for good fixation.* Keep these in mind when
evaluating cells and tissue sections.

- Fixed sample should appear similar to living sample.
- Fixation should be uniform throughout the sample.
- Cells and cellular organelles should not be swollen or shrunken.
- Proteins, lipids, or other molecules should not be washed out of cells.

There are two types of fixation, denaturing and cross-linking. *Denaturing fixation*
is not commonly thought of as fixation because it is done by heat or organic solvents.
This type of fixation destroys the molecular structure of the cellular molecules.
For proteins, denaturing breaks the 3D protein structure by breaking hydrophobic
bonds; for membranes, it dissolves the lipids into micelles. The biggest problem
with denaturing fixation is that it allows molecules to be washed out of cells during
fixation and processing. Denaturing fixatives were popular many years ago, before
modern fixatives, like paraformaldehyde, were available. The most common denaturing fixatives are cold methanol or cold acetone. These solutions are stored in
a $-20°C$ freezer and samples are submerged at this temperature for 10–20 min.
Many protocols in the literature call for these fixatives. Because of poor morphological preservation and problems retaining proteins, denaturing fixatives are not
recommended.

Cross-linking fixation is preferred. It involves chemicals with aldehyde groups that cross-link molecules within cells and tissues. Literally, the chemical fixative binds to reactive groups on proteins and lipids in the cells and holds them in the same position as if they were in living cells. Cross-linking is great for keeping the morphology in a life-like fashion, and for preventing molecules from washing out of the cells. However, the extensive cross-linking forms a molecular network that prevents antibodies from penetrating through the network and into cells and tissues.

Formaldehyde cross-links groups:

- 1° amines of Lys and Arg
- Sulfhydryl groups of Cys
- Hydroxyl groups (alcohols)
- Double bonds

Chemical Fixatives

The best fixative for light microscopic immunocytochemistry is *formaldehyde*, CH_2O (Fig. 3.1). With conditions used in tissue fixation, formaldehyde binds to amino acids, peptides, proteins, and some lipids, but not RNA, DNA, or most sugars. While formaldehyde has been shown to bind DNA (McGhee and von Hippel, 1975), in fixed tissue, most of the retention of DNA is due to protein–DNA cross-linking (Soloon and Varshavsky, 1985). Thus, even though DNA and RNA are retained in fixed tissue, the reactions with proteins occur more easily, especially with histones, which contain numerous lysine and arginine amino acids.

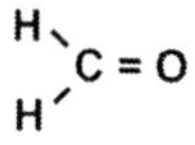

$$\begin{array}{c} H \\ \diagdown \\ C = O \\ \diagup \\ H \end{array}$$

Formaldehyde

Fig. 3.1 Formaldehyde. The chemical structure of formaldehyde shows that it is a single carbon molecule with one aldehyde

The minimum time required for fixation at room temperature is brief. From the experience with cultured cells of a few microns in thickness where retention of radio-labeled amino acids was evaluated after formaldehyde fixation, the best retention occurred at 20–30 min. The actual time for the fixation reaction has been determined to be between 3 and 5 s based on the ability of formaldehyde to stop physiological reactions (Schmiedeberg, et al., 2009). The longer times seen for amino acid retention probably involved multiple reactions that take longer. In tissue, the penetration of fixatives occurs at rates in the order of 10 mm/h, justifying fixation times of several hours for whole tissue samples. Today, where the goals of many experiments are to localize specific molecules, it is important to fix the cells or tissue long enough to insure the molecule is retained.

There are two different sources of formaldehyde, formalin and paraformalde-hyde. *Formalin* is commercially produced by oxidation of methanol and contains 37% formaldehyde and impurities including 14% methanol, small amounts of formic acid, other aldehydes, and ketones. Formalin also contains a polymer of formaldehyde – methyl hydrate polymer (Fig. 3.2). Formalin is considered a stronger fixative and is traditionally used in pathology labs for human tissue sam-ples. Because formalin contains alcohol, some molecules in cells are washed out during fixation. Also, because formalin contains additional aldehydes and ketones, proteins are fixed so that their antigens may no longer be recognized. Working fixatives are made by diluting the formalin solution 10%, resulting in a 3.7% formaldehyde concentration in the fixative. Formalin-fixed tissue has high levels of chemical autofluorescence due to the chemicals in the fixative. Formalin is not rec-ommended for cells or tissues used in research, because paraformaldehyde allows better antibody recognition of antigens.

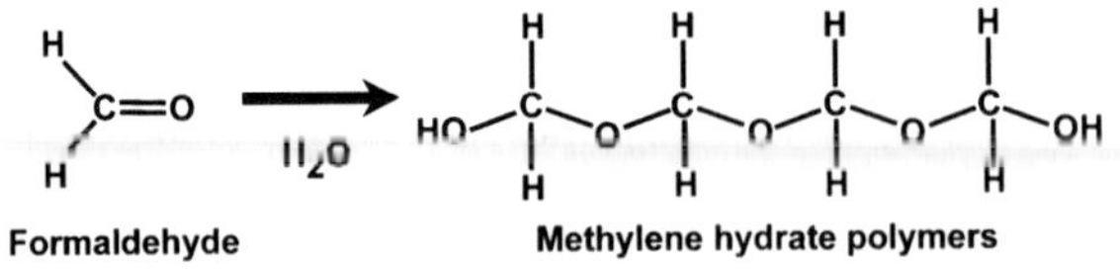

Fig. 3.2 Polymerization of formaldehyde. Over time, formaldehyde has the ability to polymerize into a methylene hydrate and can lower the concentration of formaldehyde in solution

Paraformaldehyde is a powder of polymerized formaldehyde that is made into monomers by heating and adding a chemical base, NaOH (Fig. 3.3). A solution of paraformaldehyde is pure formaldehyde and does not contain any of the impuri-ties of formalin; this is a huge advantage. The 58°C temperature used to dissolve paraformaldehyde was originally selected because at higher temperatures, some formaldehyde can become formic acid. However, at 58°C, it takes a long time to reach the endpoint of converting polymer to monomer. So people prepare the stock paraformaldehyde at higher temperatures (60–70°C) to speed the process, even with the likelihood of impurities.

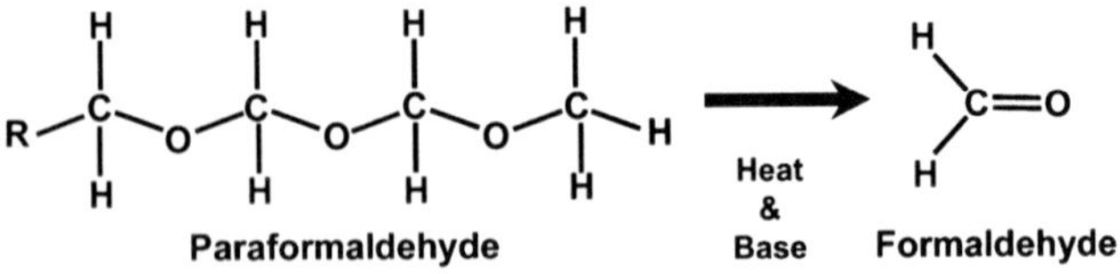

Fig. 3.3 Paraformaldehyde. To generate formaldehyde with none of the contaminates found in formalin, powder paraformaldehyde is converted to formaldehyde with heat and a small amount of NaOH

Paraformaldehyde should be used for all animal research experiments; formalin should not be used. The use of formalin with paraffin embedding requires additional steps for epitope retrieval that are not necessary with paraformaldehyde fixation. *For immunocytochemistry with fewer steps, use paraformaldehyde fixation for all animal research samples (that is, samples that are not processed for human clinical pathology).*

Disposal of formaldehyde solutions requires their collection and disposal as hazardous waste. Collect fixative and rinse solutions after fixation in a glass bottle for disposal by Environmental Health and Safety Office at your institution. Do not put formaldehyde fixatives down the drain unless the solution is deactivated with a commercially chemical treatment or unless the concentration is less than 0.1%.

Immunohistochemistry for clinical or diagnostic samples uses formalin and immunocytochemistry for research samples uses paraformaldehyde. Also, when reporting the fixative, use the term "paraformaldehyde," so that readers will know the source of the formaldehyde.

Other minor chemical fixatives – There are a variety of other fixatives that are recommended in the literature. While not recommended as the first fixative to try, several of the more common are discussed below. Each of these fixatives has potential advantages, but each also has at least one disadvantage that must be considered.

Periodate-lysine-paraformaldehyde (PLP) – It is used to increase cross-linking of molecules in the tissue to provide better fixation. Periodate oxidizes sugars attached to lipids and proteins and generates aldehydes, which bind lysine. Paraformaldehyde then cross-links the lysine. It uses 2% paraformaldehyde, 0.075 M lysine, 0.037 M sodium phosphate, and 0.01 M periodate. This method is mainly used for paraffin material to help retain cell surface components and is rarely used for non-paraffin sections. (Mclean and Nakane, 1974)

Acrolein – C_3H_4O (Fig. 3.4) (Sabatini et al., 1963) is an exceedingly fast penetrating chemical fixative that contains both an aldehyde and a double bond. Acrolein is used as a 2.5% solution along with 4% paraformaldehyde as a fixative solution. Warning – acrolein is highly poisonous, causes severe irritation to exposed skin, is extremely flammable, and is a mild carcinogen. Acrolein is NOT recommended because of its health and safety issues.

Acrolein

Fig. 3.4 Acrolein. A very effective fixative with three carbons and one aldehyde, acrolein is not used because it is difficult to work with (flammable, toxic, and carcinogenic)

Glutaraldehyde is another cross-linking chemical fixative $C_5H_8O_2$ (Fig. 3.5) (Sabbatini et al., 1963), which has a reactive aldehyde on either end of the molecule. Glutaraldehyde can form long polymers and is the single most effective cross-linking fixative. It is used only for electron microscopy because of the need for highest quality preservation. For immunocytochemistry, this fixative is too effective at cross-linking and it inhibits the diffusion of antibodies into cells and

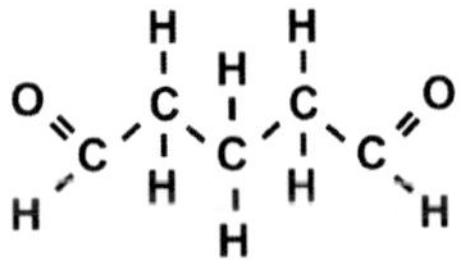

Glutaraldehyde

Fig. 3.5 Glutaraldehyde. The best cross-linking fixative with five carbons and two aldehydes, glutaraldehyde can bridge reactive groups. Not used for fluorescent immunocytochemistry glutaraldehyde is the major fixative for electron microscopy

tissues. Glutaraldehyde also generates autofluorescence, reducing the ability to distinguish fluorescent label. Thus, glutaraldehyde is not used for light microscopic immunocytochemistry.

There are several fixatives that combine denaturing and cross-linking chemical fixatives. *Bouin's* consists of 70% saturated picric acid, 10% formalin, and 5% acetic acid. This fixative is mainly used for paraffin material. *Zamboni's* is a variation of *Bouin's*. Warning – both of these fixatives are both explosive and carcinogenic. *Zenker's* is a heavy-metal fixative; the stock solution includes mercuric chloride 50.0 g, potassium dichromate 25.0 g, sodium sulfate 10.0 g, and 1000 ml distilled water. Working solution: Zenker's stock 95.0 ml and acetic acid 5.0 ml. This solution is used for paraffin material. Warning: Zenker's is both highly corrosive and carcinogenic.

Vehicle

Fixatives are made in a buffer solution or *vehicle* solution to maintain both pH and tonicity. The vehicle *maintains pH* between 7.1 and 7.3 during the fixation. Buffers are weak acids or weak bases that accept charged molecules without change in solution pH. To show the need for a buffer, examine addition of water, with no buffering capacity, as NaOH is added. The pH increases linearly and very quickly with each addition of the base (Fig. 3.6a). For a buffer used in a fixative, the addition of base at a point during the titration will not change the pH. This point is called the *pK or the pH where the buffering occurs* (Fig. 3.6b). Functionally, the buffer will allow acid or base levels to change (horizontal axis) without allowing changes in the pH (vertical axis). Buffers have pKs that range from a pH of 2 to 10, but for cells and tissue the pK range must be 7.0–7.3. The best buffer for immunocytochemistry is phosphate, but sometimes phosphate buffer reacts with calcium to form calcium phosphate precipitates. If phosphate buffer cannot be used, "Good's buffers" should be considered, specifically: MOPS p$K = 7.10$, TES p$K = 7.40$, or HEPES p$K = 7.48$. Tris with a p$K = 8.06$ is higher than physiological pH and should not be used.

Good's Buffers were selected by Dr. Norman Good because they display characteristics making them integral to research in biology and biochemistry (e.g., pK 6.0–8.0, high solubility, nontoxic, limited permeability of biological membranes) (Good et al., 1966).

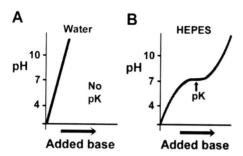

Fig. 3.6 pH buffering. For fixatives, it is important to maintain the pH at physiological levels during the chemical reactions of cross-linking in the tissue. (**a**) Titrating or adding incremental amounts of base to water results in a rapid and uniform increase in pH. (**b**) Titrating a chemical buffer, such as HEPES, has a range of pH such that adding base does not increase the pH. This range is known as the pK, and each chemical buffer will have a pK or pH at which optimal buffering capacity occurs

Buffers must *maintain the tonicity* or osmotic balance of the vehicle (Fig. 3.7). The plasma membrane lets water pass across and flow down its concentration gradient. If the solution outside the cell has the same concentration of particles as inside the cell, then the solution is *isotonic*, and the movement of water into and out of the cell is equal and there is no net change in the size of the cell (Fig. 3.7, top row). If the solution outside the cell is more concentrated (has less water) than that inside

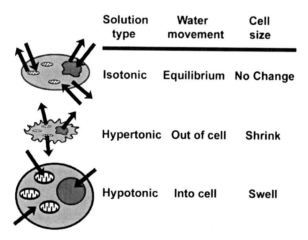

Fig. 3.7 Effects of tonicity on cells. For best fixation, the correct tonicity of the fixative solution is important. Tonicity is a measure of the concentration of particles in a solution. As a result of differences in tonicity across a membrane, water will move across the membrane to make the tonicity equal on both sides. Isotonic means that the concentration of particles is equal inside and outside the cell; here the size of the cell does not change. Hypertonic means that the concentration of particles outside the cell is higher and there is net movement of water out of a cell; the cell shrinks. Hypotonic means that the concentration of particles outside the cell is less than inside the cell and there is a net movement of water into the cell; the cell swells

the cell, it is *hypertonic*, and water flows out of the cell and into the extracellular space, causing the cell to shrink (Fig. 3.7, middle row). Or if a *hypotonic* solution outside the cell is more dilute (has less water) than inside the cell, then water flows into the cell and the cell swells (Fig. 3.7, bottom row). The same effects of osmotic force also apply to cellular organelles such as the nucleus, the mitochondria, and vesicle compartments.

In preparing fixatives, measure the concentration of particles (non-water molecules) in a solution. The units of measure here are osmoles. A reading of 300 mOsm is the normal physiological concentration inside cells, in cell culture medium, and in blood serum. When mixing a fixative, check the osmolarity before adding the chemical fixative because the chemical fixative crosses membranes and does not contribute to the tonicity of the fixative. Most labs have a pH meter and most core tissue culture labs or microscopy labs have an osmometer.

Applying Fixatives

There are several ways to apply a fixative solution. The big issue with application of the fixative is the speed once the blood supply to tissue or the culture medium of cells is removed. Cells without oxygen supply quickly begin necrotic cell death or necrosis. Once necrosis is initiated, it will spread, as lytic enzymes from membrane-bound lysosomes are released into the space around cells and begin to attack other cells. There are several methods for applying fixative depending on the sample to be fixed.

The best way to apply the fixative solution to tissue is *vascular perfusion*, where the blood vessels carry the fixative to the cells in the animal. The animal's vascular system brings the fixative solution to the cells before they have a chance to undergo necrotic cell death from lack of nutrients or oxygen. Capillaries are within microns of all the cells in the animal, so the delivery of the fixative is very rapid. This is sometimes called transcardiac perfusion and it requires equipment and preparation. This method requires some surgery skills, some specialized equipment, and must be done in a hood. Vascular perfusion is highly recommended to give the best results with animal tissue.

Another way of getting fixative solution to tissue is called *drop-in fixation*, and it is used when the entire animal cannot be perfused, such as in combined biochemical and microscopy experiments. Tissue is dissected from the live, anesthetized animal and the tissue is placed in ice-cold fixative solution. To increase access of the fixative solution to the tissue, the tissue is cut with a sharp scalpel into blocks no more than 4 mm on a side. Cutting is performed in a dish of fixative with sawing motion and minimal pressure. Deforming fresh tissue with clamps or cutting with a scissors will cause significant damage that will be seen in the microscope. Fixation of tissue blocks at 4°C with agitation should be done for a minimum of 2 h and as long as 4 h.

Fixating *cultures of attached cells* seems easy, but in reality it is difficult. The problem is that cells can float off the substrate for two reasons: (1) shooting fixative solution on the coverslip creates forces that tear off less adherent cells and (2) removing all the liquid from the cells lets them dry, which allows the surface tension to tear cells off the substrate. To determine whether either of these problems occur, look at the density of cells in the center of the coverslip on an inverted microscope before fixation and then after each step.

There are two approaches for fixing *suspended cells* or bacteria. The cells in suspension are fixed by pelleting and resuspending in solutions. In the first approach, live cells are fixed by spinning and adding 1 ml of fixative per 50 μl of packed cells. Rinse the fixed cells by spinning and resuspending for each step. For immunocytochemistry, attach cells to a slide or coverslip by letting the free cells settle on surfaces coated with 0.1 mg/ml of polylysine (rinsed four times after 30 min at room temperature before use). Alternatively, a pellet of fixed cells can be suspended in an equal volume of warm 10% gelatin and quickly pelleted by spinning. After the gelatin has hardened, the embedded pellet in gelatin is fixed again, and treated as a tissue block.

Dissecting the Area of Interest

Once fixed, the tissue must be removed from the animal. These steps involve surgical skills and instruments to find and remove the tissue. For gut or liver, the dissection involves finding the organ and cutting the tissue. Muscle, if it is to be viewed uncontracted, is held with the joints extended or stretched. For brain and spinal cord tissue, this process involves many surgical instruments and to remove the bone surrounding the tissue. Once the brain or spinal cord is removed, it needs to be further cut, which can involve the use of a vibratome or freezing microtome to achieve 50–100 μm sections. These sections are used directly or cut even smaller to include only a region of interest. For dissection of tissue from a perfused animal, it is most important to cut the tissue with minimal pressure so as not to deform the tissue.

The orientation of the tissue is usually a critical decision so that the needed plane of the section is seen in the microscope. With the 3D nature of tissue, you must decide how to orient the block of tissue so the sections provide the correct orientation. Shaping the block in a truncated pyramid with the surface for sectioning on the small square surface will always show the surface to be sectioned. For example, in kidney (Fig. 3.8), if the goal is to view the collecting ducts in cross-section then the block of tissue must be cut correctly. The medulla of the kidney should be oriented so that the surface sectioned is parallel to the outer surface of the kidney (Fig. 3.8; block on left). If the needed orientation is to view the collecting ducts longitudinally, then trim the block so that the surface to be sectioned is perpendicular to the surface of the kidney (Fig. 3.8; block on right). During the final dissection the kidney tissue block will be shaped so that it is a truncated pyramid with sections made parallel to the top of the trapezoid. This process is more efficient than trying many blocks and hoping for one to be in the correct orientation.

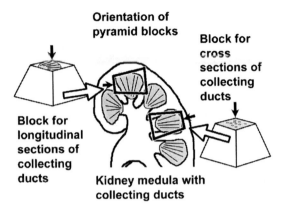

Fig. 3.8 Orientation of tissue blocks. Fixed tissue is cut into small pieces called blocks that are then sectioned. In most cases, the tissue has a specific organization so that the cells need to be held in a specific orientation. In this example, the collecting ducts of the kidney all point to the center of the kidney and they can be sectioned either as longitudinal sections or cross-sections. Blocks are trimmed into pyramids, making it easy to tell which surface is to be sectioned

Protocol – Fixation

Components for Paraformaldehyde Fixative

Preparing and using fixatives containing formaldehyde must be done in a fume hood.

1. Paraformaldehyde Stock Solution (15%); makes 200 ml

    (1) In a 500 ml Erlenmeyer beaker with 100 ml of distilled water, add 30 g of paraformaldehyde powder.
    (2) In a fume hood, place beaker on hot plate, add stir bar, and warm to 70°C.
    (3) Add drop-wise 1 M NaOH slowly and the solution will begin to clear. Add up to 1 ml or until the solution does not clear further. This solution will not be totally clear.
    (4) Let it cool to room temperature in the hood. Add 99 ml of distilled water.
    (5) Filter through the Whatman No. 1 filter paper in the hood. The solution should be clear.
    (6) Store at 4°C for no more that 1 month.

2. Phosphate buffered saline (PBS); add the following in order to a volumetric flask beginning with 800 ml of ddH$_2$O

KCl	0.2 g
KH$_2$PO$_4$	0.24 g
NaCl	8.0 g
Na$_2$HPO$_4$	1.44 g
Bring to volume	1000 ml

3. Working fixative solution 4% paraformaldehyde in PBS 100 ml

   (1) Measure out 73.3 ml of PBS into a beaker.
   (2) Add 25.7 ml paraformaldehyde stock solution.
   (3) Check that the pH is 7.3. If the pH of the fix is higher, adjust with HCl.
   (4) Use the fixative within 1 day.

4. 0.2 M Phosphate Buffer pH 7.3

   (1) Monobasic sodium phosphate (mol. wt. 137.9) 2.70 g
   (2) Dibasic sodium phosphate 7-hydrate mol. wt. (268.6) 21.45 g
   (3) H_2O distilled 500 ml
   (4) Do not adjust the pH; it should be 7.3.
       Note: Dibasic sodium phosphate is available in both anhydrous and 7-hydrate, which have very different molecular weights.

5. Working fixative solution 4% paraformaldehyde, 0.12 M phosphate buffer pH = 7.3, sucrose 60 mM; makes 500 ml

   (1) Add 300 ml of 0.2 M phosphate buffer (pH =7.3) to a beaker.
   (2) Add 10.27 g of sucrose.
   (3) Add 133.5 ml of 15% paraformaldehyde stock solution.
   (4) Add 41.5 ml of distilled water.
   (5) Stir until all of the sucrose is dissolved.
   (6) Check, the pH should be 7.3. If necessary, adjust the pH.
   (7) Use the fixative within 1 day.

Procedure

The method of whole animal perfusion is described below. Additional information is found in a review by Hoffman et al. (2008).

Perfusion Procedure

   (1) Prepare the tray with the animal support, the perfusion bottles with solutions, and the micromanipulator with a needle that has the bevel sawed off so it is not sharp.
   (2) Anesthetize the animal.
   (3) Tape animal to table with four pieces of tape, one across each limb.
   (4) Lift the sternum with notched forceps and with large scissors cut the skin at the xiphoid process.
   (5) Clamp the xiphoid process with a hemostat and cut diaphragm toward the sides of the animal. Cut the ribs on either side toward the head but stop at the clavicle.

(6) Make a small cut between the ribs at the right between ribs eight and nine to allow blood to flow out of the thoracic cage into the tray.
(7) Position the needle and turn on the flow of the saline solution.
(8) On the heart make a small cut in the right atrium to allow blood to drain. Immediately cut off the base of the heart and insert the blunt needle with the saline running. Look at the aorta to see if the tip of the needle is in place and then clamp the needle perpendicularly through the ventricles to seal the needle.
(9) After the flow from the right atrium begins to clear, or at 30 s, change the flow from NaCl solution to fixative. Turn on the fixative before turning off the saline solution. Flow of about 20 ml/min is adequate
(10) After 5–10 min, turn the fixative off and set the animal aside. The next animal should be perfused now before the first is dissected. When preparing for the next animal, be sure to run saline solution through the tubing and the needle to clear all of the fixative from the tubing.
(11) When all of the animals have been perfused dissect out the tissue of interest, and under fixative, cut the tissue into pieces.
(12) Place the pieces in the 4% paraformaldehyde fixative for an additional 2–3 h.
(13) Rinse in PBS.

Perfusion Equipment

- Anesthesia for rats
- Masking tape
- Perfusion tray
- Animal support for perfusion tray
- Micromanipulator
- Forceps
- Large scissors
- Two large hemostats
- Perfusion bottles, tubes, and needle
- Lamp
- Saline solution
- Fixative for neonatal rat 50–100 ml/animal, adult rat 200 ml/animal
- Peristaltic pump

Drop-in-Fixation

(1) Anesthetize the animal.
(2) Dissect out the tissue of interest.
(3) Place the tissue in 4% paraformaldehyde fixative; cut the tissue into small pieces.
(4) Incubate the pieces in the 4% paraformaldehyde fixative for 2–3 h with agitation.
(5) Rinse in PBS.

Chapter 4
Tissue Sectioning

Keywords Immunohistochemistry · Antibody labeling · Fluorescence micros-copy · Fluorescent immunocytochemistry · Fluorescent immunohistochem-istry · Indirect immunocytochemistry · Immunostaining

Contents

Introduction   29
Embedding Tissue by Freezing   30
Theory of Freezing Tissue   30
Freezing Tissue   32
Cryostat Sectioning   33
Tissue Processing   37
Vibratome, Freezing Microtome, and Microwave   39
Fresh Frozen Tissue   41
Embedding Tissue with Paraffin   41
Cryostat Protocol   42

Introduction

For immunocytochemistry, fixed tissue must be cut in thin sections to be viewed in the light or fluorescence microscope. There are two common ways of sec-tioning tissue – the cryostat for fixed frozen tissue and the rotary microtome for paraffin-embedded tissue. In animal research, select the sectioning method based on the experimental design. The method that gives the most reliable results and is the simplest should be selected. For immunocytochemistry, the cryostat is a very efficient and reliable method. The rotary microtome of paraffin-embedded material is more complex and problematic. For immunocytochemistry in ani-mal research, the cryostat method is recommended for reasons discussed in this chapter.

R.W. Burry, *Immunocytochemistry*, DOI 10.1007/978-1-4419-1304-3_4,
© Springer Science+Business Media, LLC 2010

Embedding Tissue by Freezing

Frozen fixed tissue is sectioned in a cryostat also known as "a microtome in a freezer." The tissue is fixed in 4% paraformaldehyde, rinsed, and then cryoprotected by *infiltration* in 20% sucrose in buffer with agitation overnight at 4°C (cold room). After 24 h the tissue blocks will sink in the solution, indicating that they are infiltrated. This infiltration step is critical. If skipped, the tissue will freeze with damaged cells and holes from ice crystals, and the resulting tissue sections on microscope slides will be brittle and might crack. The tissue must be fixed first to hold the cells in place, as cryoprotection and freezing without fixation will destroy the tissue.

Following infiltration with sucrose, the next step is freezing the tissue so that it will be held solid during the sectioning. There are a wide range of methods used for freezing tissues; we will look at the strengths and weaknesses of several of these methods.

Theory of Freezing Tissue

Freezing for morphological studies must be done very rapidly to minimize the size of the ice crystals. Any ice crystals that occur in the cells will break open the cellular organelles and the plasma membrane and destroy the morphology.

There are four basic ways of freezing tissue or cells (Fig. 4.1). The most rapid freezing, *vitrification,* which holds each water molecule in place without the presents of ice crystals (Fig. 4.1, pathway No. 1). This method is difficult to perform and involves slamming the tissue onto a silver block at liquid helium (−214°C) temperature. For light microscopic immunocytochemistry, this method is not practical.

The next method, plunging the tissue into *isopentane* (Fig. 4.1, pathway No. 2), is the best way of freezing tissue blocks for light microscopic immunocytochemistry. It is a method that features very rapid freezing and a minimum of both cost and equipment (see below for detail method). The chemical, isopentane, is a liquid at room temperature (vapor point is 26°C) and becomes solid or freezes at −160°C. Fixed infiltrated tissue is plunged into liquid isopentane near its freezing point of −160°C. The cold isopentane in contact with the tissue will have excellent heat transfer. The tissue will have only very small ice crystals in the cells and thus minimally affect organelles and will not be seen at the light microscope level. This is the preferred method for freezing tissue for immunocytochemistry.

The third method, *dry ice* (Fig. 4.1, pathway No. 3), is readily available in most labs and does not require much additional equipment. Dry ice melts at −56°C. Freezing done at this temperature is warmer than isopentane which has slower heat transfer out of the tissue. Dry ice is used as block or as pulverized with a hammer. The tissue contacts the dry ice and freezes in several seconds, which is relatively slow freezing. As such, the process generates ice crystals in the cells and can reduce the quality of the cellular detail seen in the microscope (Rosene et al., 1986). For

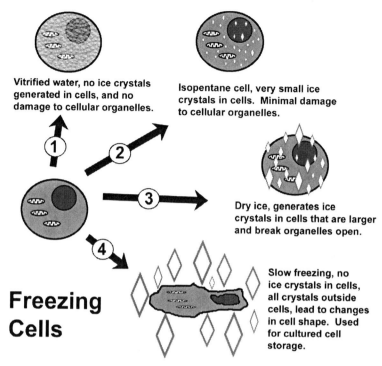

Vitrified water, no ice crystals generated in cells, and no damage to cellular organelles.

Isopentane cell, very small ice crystals in cells. Minimal damage to cellular organelles.

Dry ice, generates ice crystals in cells that are larger and break organelles open.

Slow freezing, no ice crystals in cells, all crystals outside cells, lead to changes in cell shape. Used for cultured cell storage.

Freezing Cells

Fig. 4.1 Freezing cells. To prepare for sectioning in a cryostat, tissue infiltrated with 20% sucrose is frozen. The rate of freezing will determine size of the ice crystals and the quality of the cellular morphology. With smaller ice crystals, less disruption of tissue and cellular morphology is seen. 1. The most rapid freezing at –214°C gives the best morphology and water becomes vitrified; no ice crystals are found. 2. Rapid freezing in isopentane at –160°C generates small ice crystals; damage is not seen in light microscopy. 3. Dry ice freezes at –56°C and generates ice crystals large enough to show cellular damage and holes. 4. Slow freezing generates large ice crystals outside cells with major disruption of cellular morphology

labs that are looking at cell death, where cell relationships to each other are important, dry ice can introduce cellular damage and cracks in tissue. The problem here is that such cracks might not be interpreted as artifacts of sample preparation.

A variation of the dry ice method is to make a slurry of dry ice in ethanol and obtain a slush that is colder near –75°C. Tissue exposed to this slush is frozen more quickly than with dry ice alone. However, the tissue now has a layer of frozen alcohol that will infiltrate the tissue when it is thawed. Exposing the tissue to ethanol will change its interaction with antibodies and other processing reagents.

The fourth method, *slow freezing* (Fig. 4.1, pathway No. 4) should not be used for morphology but rather only for preserving living tissue culture cells for thawing later. With very slow temperature loss rates of tenths of a degree per minute, ice crystals form so slowly that all ice crystals are outside of the cell. The slow freezing is done over a 12-h period and all the ice crystals that form are extracellular with the morphology of the cell "pushed around" by the ice crystal formation. With only extracellular ice, the cells will be deformed, organelles changed, and extracellular spaces enlarged. This method is ideal for freezing cultured cells to be stored in liquid nitrogen, because it does not have intracellular ice crystals that would destroy cellular organelles.

The term *Snap Freezing* is widely used to describe freezing tissue for immunocytochemistry. The term comes from food preparation industry, where rapid freezing is used. For biomedical sciences snap freezing has no single definition and can mean freezing in liquid nitrogen, on dry ice, or in isopentane. Freezing tissue directly in liquid nitrogen is very slow because the vapor temperature (change from liquid to gas) is –196°C. The tissue plunged into liquid nitrogen is surrounded by a shell of nitrogen gas until the tissue cools to –196°C, when it can join the liquid phase. For tissues held at room temperature or slightly below, liquid nitrogen is a poor freezing agent! For the sake of clarity avoid the term "snap freezing," instead name the actual agent and method.

Freezing Tissue

Think about freezing methods in terms of the tissue samples and the desired outcome. Initially, determine the size of the tissue block, the number of tissue blocks to be sectioned at one time, and the requirements for specific orientation of the tissue. Because most scientists need a particular orientation of their section, the tissue must be frozen in a specific position to give the needed sections.

Once infiltrated with sucrose, the tissue block is frozen by surrounding it in a liquid, which holds the frozen tissue for sectioning. The most common liquid is "optimum cutting temperature" or *O.C.T.* (Sakura Tissue-Tek); another embedding liquid is tissue freezing medium or TFM (Triangle Biomedical Sciences).

Isopentane (Sigma M32631) is the best freezing agent for sucrose-infiltrated tissue for immunocytochemistry. In a plastic beaker that is cooled in liquid nitrogen, add the liquid isopentane (Fig. 4.2). If liquid nitrogen is not available, a dry ice alcohol slurry works, but it is not as cold. Isopentane will freeze solid white in liquid nitrogen, so to thaw frozen isopentane push a piece of metal, such as the blade of a screw driver, into the isopentane until liquid.

A preferred method for freezing samples is to use *strips of aluminum foil* (7- to 8-mm wide and 40-mm long) with an identification pressed in at one end with a pencil and the sample sitting on the other. Remove the tissue from the 20% sucrose solution, blot on a piece of paper towel, and set on the other end of the foil. Immediately with a forceps, plunge the foil strips with the tissue into the cold isopentane (Fig. 4.2). The tissue will stick to the foil and it can be placed in a vial in the liquid nitrogen bath and is subsequently stored in a –70°C freezer in individual small tubes. For sectioning, the frozen tissue is mounted later in O.C.T. on a chuck.

Another method for freezing samples uses *plastic molds*. There are many varieties of small disposable plastic molds for one or more pieces of tissue to be positioned next to one another. Rather than freezing the tissue blocks individually

Fig. 4.2 Method of freezing
tissue. To freeze tissue
rapidly, the amount of
material frozen needs to be as
small as possible. Freezing
tissue with no liquid is best.
On a strip of aluminum foil,
write the sample information,
place a blotted price of tissue
on the foil, and plunge in
isopentane. Freezing tissue
surrounded by liquid has a
slower rate of freezing, but
this method is sometimes
used. Place a piece of tissue
in a mold, cover with O.C.T.,
and plunge in isopentane

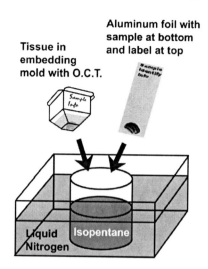

in molds, several blocks of tissue are grouped in one mold and frozen together. This reduces the number of blocks that need to be sectioned. Prior to freezing, when the mold is at room temperature, use a permanent felt tip pen to write the sample information. Place O.C.T. in the mold and arrange the tissue blocks in the mold. Be sure to orient the side of the block to be sectioned to the bottom of the mold. Also, note the organization of the blocks so that they can be identified after sectioning. When the mold is ready, use a forceps to plunge it into the cold isopentane. The frozen mold with tissue can be held in the liquid nitrogen bath until transfer to a –70°C freezer (Fig. 4.2). Plunging a mold into isopentane freezes it within a second, faster than dry ice but not as rapid as the aluminum foil.

The simplest method for freezing tissue is on dry ice. This method is not ideal and can lead to tissue shrinking and should be used if dry ice is the only available option. Blot off most of the sucrose solution and place the infiltrated tissue in small aluminum foil packages with pencil writing as labels. Place these foil packages in crushed dry ice to allow freezing. The frozen tissue in the foil packages can be stored in a –70°C freezer until sectioning.

Cryostat Sectioning

A *cryostat* is microtome in a freezer and is used to cut frozen sections. The initial steps in using a cryostat are to put the frozen samples on a "chuck" that is used for sectioning.

The metal platform, called a "chuck" (Fig. 4.3), holds the tissue on the microtome arm for sectioning. There are many types of chucks, but all have ridges or grooves to hold the frozen tissue during sectioning. Frozen tissue blocks are attached to a chuck just before sectioning, and following the sectioning, the block is removed and the

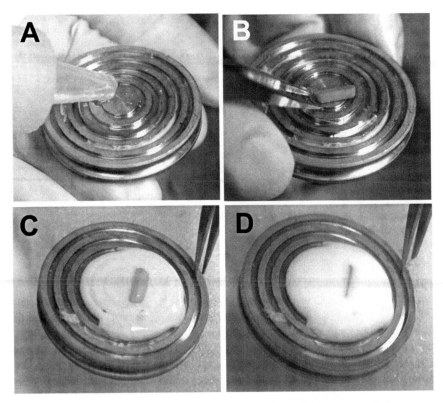

Fig. 4.3 Mounting frozen tissue on a chuck. Tissue is held for sectioning on a metal chuck frozen in O.C.T. (**a**) Room temperature O.C.T. (clear) is applied to a metal chuck. (**b**) The tissue is placed in the O.C.T. with a forceps and the chuck is placed in the cryostat or on dry ice. (**c**) The O.C.T. begins to freeze; note that the O.C.T. turns white with the tissue in the center. D. When fully frozen, the white O.C.T. surrounds the tissue

chuck is cleaned. Tissue blocks are not stored on chucks. Chucks are not stored in the cold cryostat because they build up a layer of frost that does not allow the tissue to freeze cleanly to the chuck. Begin each session of freezing with chucks at room temperature and freeze each chuck just before use.

Tissue that has been previously frozen on aluminum foil and stored in −70°C freezer is kept cold on dry ice and can be picked up with a forceps. The chuck is cooled in the cryostat or on dry ice. Quickly, place O.C.T. on the cold chuck (Fig. 4.3a), submerge the frozen tissue in the freezing O.C.T. (Fig. 4.3b). As the O.C.T. freezes, it changes from clear to white (Fig. 4.3c, d). To shorten the amount of time needed for sectioning, leave the block near the top surface of the O.C.T. (Fig. 4.3d).

Tissue that has been frozen in molds is frozen to the chuck with a small amount O.C.T. Remove the frozen block with the tissue by warming the sides of the mold. The chuck is cooled in the cryostat or on dry ice. Quickly, place a layer of O.C.T. on the cold chuck, and push the bottom of the block against the O.C.T. on the chuck. When frozen, the O.C.T. bonds the block to the chuck.

Prepare the cryostat and put the chucks with the tissue in the cryostat freezing tray to equilibrate with the temperature of the cryostat (Fig. 4.4a, *arrow*). Attach the chuck to the cryostat arm and tighten to hold it firmly (Fig. 4.4b). Watch for frost on the shaft of the chuck because it might not let the holder tighten firmly.

Trim the top, bottom, and sides so that a small amount of frozen O.C.T. is surrounding the tissue. With a single edge razor blade, shave off a couple of millimeter at a time and cut perpendicularly to remove the shavings if necessary (Fig. 4.4c, d). Do not trim the face of the block with the razor blade.

Adjust the knife position from front to back (Fig. 4.4e). Trim the excess frozen O.C.T. on the block face by cutting thick sections until the tissue begins to appear in the sections. Once the blade starts to cut into the tissue, adjust the section thickness using the micrometer left-hand side of the cryostat to the thickness that is appropriate. Remove pieces of tissue and ice from the knife edge by brushing firmly upward with a small, stiff-,bristle paint brush.

Alignment of the anti-roll plate is important to obtaining good sections. Adjust the handle on the anti-roll plate back into position. It should be exactly parallel to the knife edge. As sections are cut, they will slide under the anti-roll plate keeping them flat (Fig. 4.4f). Lift the plate and use a small paint brush to handle the section. To keep the section flat, use the brush to press the section down on the knife. Some people section without the anti-roll plate and use the paint brush to keep the sections flat on the knife (Fig. 4.4 g).

Mount sections on microscope slides for processing with antibodies. Cryostat sections can be collected on microscope slides; SUPERFROST® PLUS slides hold sections well. Alternatively, glass microscope slides can be coated with a 0.1 mg/ml poly-L-lysine (Sigma P1399) solution in water. Loss of sections during processing is a frequent problem.

To pick up the sections, have a slide ready at room temperature. If used, rotate the anti-roll plate out of the way. Orient the slide with the bottom of slide closest to the knife (Fig. 4.4 h). Slowly lower the top of the slide to touch the frozen section and it will thaw on the microscope slide (Fig. 4.4i). Additional sections can be put on one slide if needed. Usually more than one section or even a ribbon of sections is cut for each slide. This increases the chances of having at least one good section per slide. Most importantly, this reduces the number of slides that need to be processed for immunocytochemistry. Finally, remove the O.C.T. from around the sections. With a fine pair of forceps (EM 3C or EM 5) pick an edge of the dried O.C.T. and pull it down the slide to the opposite end of the section (Fig. 4.4j). The sections will remain on the slide and the thin transparent O.C.T. will peel off. In Fig. 4.4j, the section indicated by the *white arrows* still has dried O.C.T., while the small circular sections to the bottom have the O.C.T. removed. Sections on slides can be stored in a −20°C frost-free freezer for an extended period of time. Sections stored for several years have been thawed and used successfully.

When done sectioning, remove all tissue waste from the machine tissue. O.C.T. left in the cryostat will turn into a sticky mess if not removed after each use. Clean

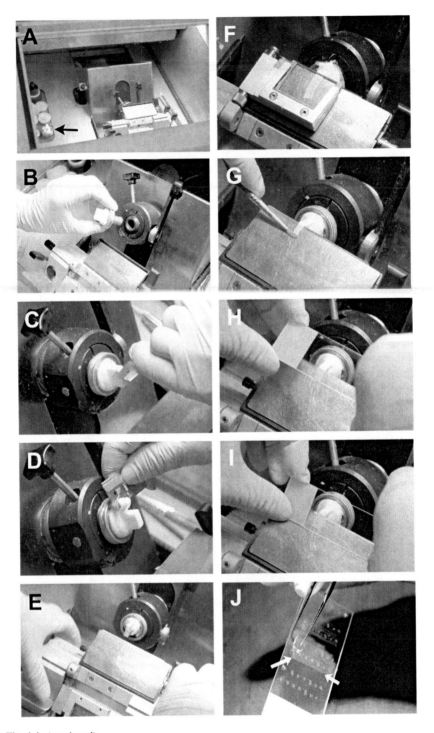

Fig. 4.4 (continued)

all exposed surfaces with alcohol. Reset the advance mechanism to the top of its travel for the next user. Turn off the light. Temperature settings should be left where they were set for sectioning. Close the door completely.

Tissue Processing

Cryostat tissue sections on microscope slides are now used for immunocytochemistry, but first they need a barrier to confine the antibody solutions. Use a *hydrophobic pen* to outline the sections with a box of hydrophobic material (Fig. 4.5a, b) from a PAP-Pen (EMS, Hatfield, PA) or an ImmunoPen (most biomedical supply companies), which will repel water and keep the incubation solutions in the box. While good in retaining the incubations solutions, use of these pens to create the barrier lines requires care. Apply the pen only to totally dry glass microscope slides that are clean and free of dried buffer or dried tissue so the pen lines will attach. The barriers made with the pen need to be completely dry (at least 2 h and generally more) before use or they will detach. During processing for immunocytochemistry, do not touch the barrier lines with pipette tips and do not attempt to dry the barrier lines if they become wet; the solution will naturally bead off the barrier. The amount of incubation solution put in the box defined by the hydrophobic barrier needs to be small enough so that it will form a small bubble and stay in place when moved (Fig. 4.5c). It is important to determine this volume for the size boxes used and most often between 35 and 100 μl is sufficient.

Incubations of microscope slides with small amount of incubation solution must be done in a closed box with a source of water vapor (moist paper towels) to prevent the small volume of solution from drying. While it is possible to buy boxes, most researchers like to make their own with recycled plastic boxes from lab products. Incubation times are generally 24–48 h. Evaporation of the small volumes is a major problem. Agitation, while needed to shorten the incubation time, is not used because of the danger that the incubation solution will run outside the rings.

Fig. 4.4 Cryostat. Use a cryostat to cut sections from tissue frozen on a chuck. (**a**) The chamber of a cryostat with the frozen tissue indicated by the *black arrow*. This tissue was frozen in a rectangular mold. (**b**) The chuck with the tissue is placed in the arm. (**c**) The excess O.C.T. is trimmed off with a razor blade by cutting into the block several times. (**d**). The trimmed frozen O.C.T. is removed with a razor blade cut from the *top* of the block. (**e**) The knife is adjusted into position very close to the block, but not touching. The block is advanced toward the block until sections are cut. (**f**) The anti-roll plate prevents cut sections from curling. (**g**) An alternative to the anti-roll plate is to use a fine paint brush to hold a section as it comes off the block. (**h**) Sections are picked up on microscopes slides. (**i**) Lower the *bottom* of the slide until the section attaches; then warm the slide. (**j**) When the section is dry, remove the dried film of O.C.T. (*arrows*) with a forceps. Note the two rows of sections toward the slide label (*down*) have the film already removed

Fig. 4.5 Incubating tissue sections. For incubations of tissue sections on microscope slides, wells are made around sections. (**a**) To hold the incubations solutions in wells around sections, a pen with hydrophobic material is used to paint a rectangle around the section. (**b**) Section through the microscope slide showing the line made by the pen and the tissue section. (**c**) The incubation solutions are held in place by the hydrophobic lines

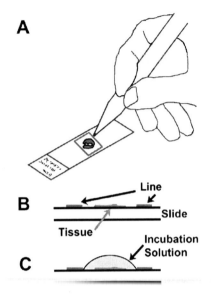

Cultured cells can be grown on glass coverslips or in Lab-Tek chamber slides, depending primarily on the growth requirements of the cells. Cultured cells on 12 mm coverslips coated with a 0.1 mg/ml poly-L-lysine can be easily placed in a 24-well culture plate with a Dumont 3C forceps (Fig. 4.6a) for processing. Each well is a container for one coverslip and contains the incubation solutions (Fig. 4.6a). To cover the sections in wells, use 250 μl per well of incubation solution (Fig. 4.6b) so that the meniscus will cover the section. Incubation times for cultures in 24-well plates on a rocker are 3–4 h for the primary antibody and 1 h for the secondary antibody. A rocker for the 24-well plate will speed incubations. Because the cells are in a thin layer and are agitated, the incubation times are substantially shorter than for tissue sections. Removing coverslips from the well requires practice with a Dumont 3C forceps and a long syringe needle bent at a right angle. Always hold the coverslip in 3C forceps so that the cells are on the side with the tape label on one side of the forceps. Alternatively for incubations, coverslips can be processed on small drops (30–100 μl) of reagents. The drop is placed on a piece of Parafilm and the coverslip is inverted on the drop.

Chamber slides are best for cells that require tissue culture grade plastic as a substrate. Lab-Tek chamber slides come with such a slide and they come in a range of well sizes, but a very practical size is the 4-well plate (Fig. 4.7). The microscope slide base is made of Permanox, the same plastic used for plastic petri dish bottoms so that the cells grow like they do on petri dishes or flasks. The nice thing about the chamber slides is that wells on the slide do not need be removed until after immunocytochemistry processing. The wells can be removed just before mounting a

Fig. 4.6 Incubating cultures
on coverslips. Cell cultures
on 12 mm coverslips are
incubated in wells of a
24-well plate. (**a**) Individual
12 mm coverslips can be
placed in wells. (**b**) The
coverslips can be incubated in
250 μl of solution per well

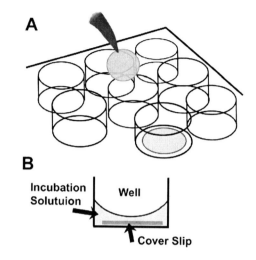

Fig. 4.7 Chamber slides for
cell cultures. To grow
cultured cells on a tissue
culture plastic substrate for
microscopy, one choice is
chamber slides. The
individual wells can be used
to plate cells and process cells
for immunocytochemistry.
The *upper chambers* are
removed along with a gasket
from the plastic slide; the
slide can be covered with a 22
× 50 mm coverslip

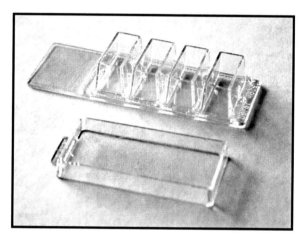

single 50 × 22 mm glass coverslip. For Permanox slides, use only aqueous mounting
medium (see below) and not organic based (e.g., Permount).

Vibratome, Freezing Microtome, and Microwave

A *vibratome* is used to cut delicate fixed animal tissue and to prepare it for sectioning
in a cryostat. This device works best with soft fatty tissue like brain or breast tissue
to give sections from 25 to 500 μm. The tissue is held in place without infiltration
by surrounding it with an agar. The tissue is attached to a plate in the vibratome
with superglue and then covered with low melting point agar before the superglue

hardens. The agar block is next submerged in a buffer bath and the cutting occurs under buffer. The vibratome uses a razor blade in a vibrating head to cut tissue surrounded by agar, which is then submerged in buffer. Sections are handled with small paint brushes and placed in buffer in a 24-well tissue culture plate. After the vibratome sections are collected, they are placed on a flat surface under buffer and the agar surrounding the tissue is dissected off. These sections can be used for processing as free-floating section not attached to coverslips or microscope slides, or they can be infiltrated with sucrose and sectioned in a cryostat.

A *freezing microtome* (Fig. 4.8) is used for cutting 15–50-μm thick sections of fixed tissue. The fixed tissue is infiltrated with 20% sucrose placed on the chuck and quickly frozen with liquid CO_2. One advantage of this device is its ability to easily and quickly cut thicker sections than the cryostat. These sections are incubated for immunocytochemistry as floating sections. The freezing microtome is most commonly used for brain tissue, where thick sections reduce the number of sections processed for immunocytochemistry and sequential sectioning results. This freezing method has the potential to generate artifacts.

Fig. 4.8 Freezing microtome. To cut thick frozen sections of 15–50 μm, a freezing microtome will do the job. The cryoprotected tissue is frozen in O.C.T. on the chuck with liquid CO_2, a knife is brought across the tissue, and the section is collected with a small paint brush and placed in a tray

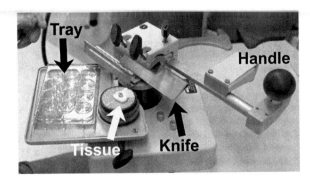

Floating section immunocytochemistry can be performed on free floating, thick sections (25–100 mm) cut on a vibratome or freezing microtome. The advantage is that antibodies penetrate deeper because of the greater movement of the section floating in the solutions. Also both sides of the section will be exposed to the antibodies. Sections thicker than 30 μm will have a gradient of decreased labeling toward the center of the section. Floating sections are more common for enzyme label described in Chapter 6, Labels. Take care when mounting these very thick sections to avoid flattening of the tissue with the coverslip. The protocol for floating sections is the same as for attached sections, except that the 1° antibody incubation is extended.

A new technique uses microwaves to speed the diffusion of reagents into tissues. *Microwaves* can be used for ultra-rapid processing of tissue and speeding incubation times. While the theories are still not well understood, they might go something like this: Microwaves cause increased vibration of polarized molecules in the tissue and this speeds the diffusion. The microwave has a cooling bath so that the tissue samples stay at room temperature during the exposure. This method is useful for very rapid initial fixation, deep penetration of chemicals and antibodies, or rapid tissue processing. An example of a commercial product is PELCO BioWave®, which allows control of many parameters and shortens incubations of hours to minutes.

Fresh Frozen Tissue

The *fresh frozen tissue* from a human surgical biopsy is frozen without fixation or cryoprotection for rapid diagnosis by pathologists when a patient is still on the operating table. However, the pathologist who looks at these fresh frozen sections requires paraffin sections from the same sample to later confirm the initial observations. In fresh frozen tissue, the morphology is distorted and because no cross-linking fixative has been used, components of the tissue can wash out of the tissue during processing. *Fresh frozen tissue should never be used for immunocytochemistry with animal research.* Always fix tissue before any other procedure.

Embedding Tissue with Paraffin

Human tissue should be fixed in 10% neutral buffered formalin and then dehydrated for embedding in paraffin. Paraffin is nonaqueous embedding medium, so the tissue blocks must have the water removed or be dehydrated. Dehydration is done in organic solvents such as alcohol, acetone, xylene, or toluene. After dehydration, the tissue blocks are embedded with liquid (warm) paraffin. When cooled, the wax embedded block is sectioned on a rotary microtome. Before immunocytochemistry can be performed on the resulting tissue sections, they must be rehydrated by processing with the same organic solvents back to water. Thus, the dehydration and rehydration steps are needed before immunohistochemistry.

Performing immunohistochemistry on these rehydrated paraffin sections frequently leads to poor results. In contrast, the same antibodies on paraformaldehyde-fixed and cryostat sections will give good results. The issue with formalin-fixed and paraffin-embedded tissue is that the exposure to formalin and dehydration alters the epitopes in the tissue. As a result, formalin-fixed and paraffin-embedded tissues need additional processing methods, known as *epitope retrieval or antigen retrieval*. Done before immunohistochemistry, epitope retrieval involves heating the sections in buffer with either an acid or base to allow the antibody to recognize the epitope. Also, the exact process of epitope retrieval can be different for individual antibodies. There are numerous papers and books on epitope retrieval and how to apply the method.

Traditionally, formalin-fixed and paraffin-embedded tissue is used by pathologists for examining human tissue. This method was first adapted for immunocytochemistry with tissue from research animals. Today, for animal research, there is no requirement to use formalin-fixed and paraffin-embedded tissue methods. Table 4.1 shows the steps necessary for formalin/paraffin immunohistochemistry

Table 4.1 Comparison of fixation and embedding methods

Steps	Formalin/paraffin	Para/cryostat
Fixation	X	X
Dehydration	X	
Embedding/freezing	X	X
Sectioning	X	X
Rehydration	X	
Epitope retrieval	X	
Antibody incubations	X	X
Microscopy	X	X

and the steps for paraformaldehyde/cryostat immunocytochemistry. The X indicates a required step and the blank indicates steps not used in the method. The paraformaldehyde/cryostat method has three fewer steps than the formalin/paraffin method, which alone is justification for using the paraformaldehyde/cryostat method in animal research. In addition, the formalin/paraffin method is generally performed in a histology core lab and the paraformaldehyde/cryostat method can be done in your lab. Finally, paraformaldehyde fixation and cryostat sectioning is more flexible with new antibodies used with the steps described in these chapters, while pathology labs generally have a limited set of antibodies they will use.

Cryostat Protocol

(1) Remove the fixed and rinsed tissue from the animal and cut into cubes called blocks. Perform this trimming under buffer with a new scalpel blade and sawing motion with light pressure. The tissue should not deform from the pressure of the scalpel blade. Blocks should be 5–10 mm on a side. Go to the next step immediately, or hold the tissue at 4°C in buffer for upto week.

(2) Immerse blocks in 20% sucrose in buffer until they sink. This requires incubation overnight at 4°C with agitation (cold room on a shaker) or until the tissue sinks. Hold infiltrated blocks at 4°C for no more than a few days.

(3) Prepare blocks using one of the following methods:

(a) Make strips of heavy-duty aluminum foil and write the sample information on each strip of foil. Just before freezing, remove the tissue block from the sucrose solution, blot to remove excess liquid, and place on other end of foil strip opposite from the information and freeze.

(b) Fill a mold with enough O.C.T. to just cover the tissue. Place the pieces of tissue in the mold with correct orientation and freeze.

(4) Freeze the tissue by one of the following methods:

(a) In a 50 ml plastic beaker, pour 20 ml of isopentane and place the beaker in styrofoam container with liquid nitrogen (if liquid nitrogen is not available, a dry ice–methanol slurry can be used). If the isopentane freezes solid, thaw the isopentane with the blade of a screw driver, and remember to touch only the insulated screw driver handle and not the cold blade. Plunge the tissue into the cold liquid isopentane and hold under liquid for 5–10 s to be sure that all of the O.C.T. has turned white. Following freezing, store individually in molds at –70°C.

(b) Block dry ice is best used by pressing the foil of mold onto the solid ice directly. Following freezing, store individually in small plastic tubes at –70°C.

(5) Prepare the cryostat. Turn on the light. The temperature of the cryostat chamber is at –20°C and that the quick freeze compartment is at least –40°C. Check the position of the advance mechanism so that it is reset to the farthest-back position. This will vary depending on the use of a large metal knife or a small disposable blade in a knife-shaped holder. Place a blade into the holder if needed. Check to confirm that the knife angle is set correctly. Accumulate the necessary brushes and microscope slides.

(6) Before sectioning, mount the frozen tissue block on the chuck. Remove the frozen blocks from the –70°C freezer and arrange them on the chuck for sectioning. With a chuck on dry ice, or the cryostat cabinet, before it freezes correctly orient the tissue for sectioning. In the case of tissue frozen in molds, the contents of an entire mold can be removed from the mold with gentle warming. Place the frozen tissue in frozen O.C.T. on the chuck on dry ice or in the cryostat to freeze. Place the chucks with the tissue in the cryostat freezing tray to equilibrate with the temperature of the cryostat. Always carry your tissue samples on dry ice. Attach the chuck to the cryostat arm and tighten to hold it firmly. Watch for frost on the shaft of the chuck because it might not let the holder tighten firmly.

(7) Align the tissue so that the side to be sectioned, block face, is parallel to the knife edge. Loosen the knob and tilt the chuck side to side or top to bottom as necessary. This will help to achieve even cutting of the block and prevent sections from crumpling.

(8) Trim the top, bottom, and sides so that a small amount of frozen O.C.T. is surrounding the tissue. With a single edge razor blade, shave off a couple of mm at a time and cut perpendicularly to remove the shavings if necessary. Do not trim the face of the block with the razor blade.

(9) Adjust the knife position front to back. Trim the excess frozen O.C.T. on the block face by turning the large crank wheel on the right-hand side of the cryostat counterclockwise. Cut thick sections until the tissue begins to appear in sections. Some cryostats have a "trim" mode that allows for very thick sections to be cut while advancing to the tissue. If the cryostat does not have a trim mode, initially set the thickness of the sections to 20 µm and once the blade starts to cut into the section, adjust the section thickness using the micrometer left-hand side of the cryostat to the thickness that is appropriate. Remove pieces of tissue and ice from the knife edge by brushing firmly up with a small, stiff-bristle paint brush.

(10) Align the anti-roll plate to ensure obtaining good sections. Adjust the handle on the anti-roll plate back into position. The edge of the anti-roll plate should be exactly parallel to the knife edge and it should also line up with the knife edge. As sections are cut, they will slide under the anti-roll plate, keeping them flat. Lift the plate and use the brush to press the section down on the knife. In many applications, it is not necessary to use the anti-roll plate but instead use a paint brush to keep the sections flat on the knife.

(11) Adjust the sectioning temperature from −20 to −15°C or −10°C for harder tissue. Adjust the speed of sectioning. Generally, harder tissues cut best at faster speeds and softer tissues at slower speeds.

(12) To pick up the sections have a slide ready at room temperature. Frequently, set the microscope slides on the open glass door of the cryostat. Leave the microtome arm at the bottom of its travel. If used, rotate the anti-roll plate out of the way. Orient the slide with the bottom of slide closest to the knife. Slowly lower the top of the slide to touch the frozen section. The section will jump onto the microscope slide and it will thaw on the slide. Additional sections can be put on one slide if needed. Usually more than one section or even a ribbon of sections is cut for each slide, reducing the number of slides that need to be processed for immunocytochemistry.

(13) Remove the O.C.T. from around the tissue. With a fine pair of forceps (EM 3C or EM 5), pick an edge of the dried O.C.T. and pull it down the slide to the opposite end of the section. The tissue will remain on the slide and the thin transparent O.C.T. will peel off.

(14) When finished sectioning, remove all tissue waste from the machine tissue. O.C.T. left in the cryostat will turn into a sticky mess if not removed after each use. Clean all exposed surfaces with alcohol. Reset the advance mechanism to the top of its travel for the next user. Turn off the light. Temperature settings should be left where they were set for sectioning; close the door completely.

Chapter 5
Blocking and Permeability

Keywords Immunohistochemistry · Antibody labeling · Fluorescence micro-scopy · Fluorescent immunocytochemistry · Fluorescent immunohistochem-istry · Indirect immunocytochemistry · Immunostaining

Contents

Introduction   45
Nonspecific Antibody Binding to Tissue and Cells   45
Blocking for Nonspecific Antibody Binding   47
Permeabilize Tissue and Cells to Allow Antibody Penetration   49
Effects of Blocking Agents on Antibody Penetration   51
Combined Incubation Step   53

Introduction

There are two key elements in immunocytochemistry: (1) *to allow binding of the antibodies only to appropriate sites and* (2) *to allow antibodies access to antigens inside cells.* Antibodies are proteins and can bind to tissue in ways other than specific via Fab binding. If the antibodies bind nonspecifically, then the labeling will occur nonspecifically. Blocking of these nonspecific sites is an important and necessary step. A second requirement is that tissues and cells must be treated to allow antibodies to get inside the membranes to the cytoplasm. This is called permeabilization of cells and is done with detergents. Blocking and permeabilization are combined in a single incubation step in the immunocytochemistry protocol after fixation and before the antibody incubation.

Nonspecific Antibody Binding to Tissue and Cells

Antibodies can bind to cells in a highly specific fashion based on "Fab-epitope binding." However, antibodies can bind to cells by nonspecific means that will give incorrect results.

R.W. Burry, *Immunocytochemistry*, DOI 10.1007/978-1-4419-1304-3_5,
© Springer Science+Business Media, LLC 2010

In tissues and cells there are four types of sites where an antibody can bind (Fig. 5.1), but only one is the correct site. *Nonspecific binding occurs when an antibody binds by a mechanism other than by Fab-epitope binding.*

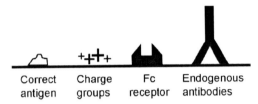

Correct   Charge   Fc   Endogenous
antigen   groups   receptor   antibodies

Fig. 5.1 Sources of nonspecific binding. Incubation with antibodies should have only antibody binding to the correct antigen. Nonspecific binding can result from charged groups that bind proteins including antibodies. The tissue can have Fc receptors that bind to the Fc region of any antibody. Some tissues can have exposed antibodies

The correct antigen (Fig. 5.1) will bind only to its specific antibody. The first type of nonspecific-binding site is charged groups (Fig. 5.1), which are naturally present in cells and can be added during the process of chemical fixation. A second type of nonspecific-binding site is Fc receptors (Fig. 5.1), which occur on circulating white cells, macrophages, and natural killer cells. These cells are rarely seen in normal tissues, but are present at high concentrations at inflammation sites. Endogenous antibodies (Fig. 5.1) are seen when an antibody produced in the living animal binds to an antigen. This is most common in inflammation where antibodies bind invading bacteria.

The difficulties that arise from nonspecific binding of incubated antibodies to charged groups, Fc receptors, and endogenous antibodies are major problems for immunocytochemistry (Fig. 5.2).

Correct Antigen contains the epitope used to generate the antibody.

Charged groups are the most common nonspecific sites for binding of antibodies (Fig. 5.2). Charged groups are normal on proteins or lipids and sometimes the result

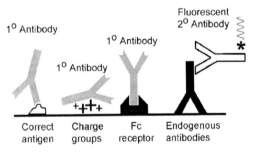

Correct   Charge   Fc   Endogenous
antigen   groups   receptor   antibodies

Fig. 5.2 Examples of nonspecific antibody binding. During immunocytochemistry incubations nonspecific binding occurs. The correct antigen will bind the primary antibody correctly. Charged groups will bind to any protein or antibody such as the 1° antibody. The Fc receptor will bind the first antibody that comes in contact here with the 1° antibody. Any endogenous antibodies are likely to bind only to secondary antibodies made against the species of the animal

from the fixation process. Histones contain numerous charged basic amino acids and can be a source of nonspecific nuclear binding. Use of formalin or glutaraldehyde fixatives will result in charge groups with nonspecific labeling, which is diffused throughout the tissue section and not localized to specific parts of the tissue section or the cultured cells.

Fc receptors in tissue bind the Fc end of the primary antibody and lead to incorrect labeling (Fig. 5.2). Fc receptors are found on macrophages and other immune cells, and can bind any antibody in the immunocytochemistry procedure. The primary antibody would most likely bind open Fc receptors as the first antibody incubation. This problem occurs only in areas where there are macrophages, such as injury and inflammation sites. Unless cell cultures include macrophage or natural killer cell lines, Fc receptor binding does not occur in cultures.

Endogenous antibodies in tissue are only a problem for 2° antibodies for that species of animals (Fig. 5.2). This situation is only a problem when the species of the tissue is the same as the species of the primary antibody, such as using a mouse antibody in mouse tissue or a rabbit antibody in rabbit tissue. Such endogenous antibodies will bind secondary antibodies of the immunocytochemistry procedure. For example, when using a mouse primary antibody and a labeled goat anti-mouse fluorescent secondary, the potential exists for the secondary to bind to the endogenous mouse antibodies in the tissue. Practically, this problem is only seen in animals where there is an injury or inflammation and the antibodies are seen at the inflammatory site. In these cases, the labeling is inconsistent across the entire section and is highest over the site of inflammation. In cultured cells, this is not a problem except when using specific immune system cell types.

Blocking for Nonspecific Antibody Binding

To prevent the inappropriate binding of antibody to nonspecific binding, a series of blocking agents is needed. Each of the three nonspecific binding sites requires a different blocking agent (Fig. 5.3). The correct binding of the specific 1° antibody to the antigen is not effected by the blocking agents.

For charged groups, the most common blocking agent is *bovine serum albumen* (BSA) at 10 mg/ml, because albumin is nonantigenic and will bind charged sites (Fig. 5.3). BSA then acts as a nonspecific blocking agent to cover any charge groups and will not bind to antibodies. Be sure to purchase BSA that is free of IgG molecules, namely purified BSA that has been fractionated. Many molecular genetics procedures use BSA, but do not have the specificity for the presence, even in trace amounts, of IgG. One of the least costly fractionated BSAs is BSA fraction V, which is recommended. Normal serum includes albumin which also blocks charged groups. Sometimes, normal serum from the species of the animal used for the 2° antibody is used alone without BSA.

For example, when using a mouse monoclonal antibody against tubulin in rabbit tissue with a secondary goat anti-mouse 488 fluorophore, the blocking step would

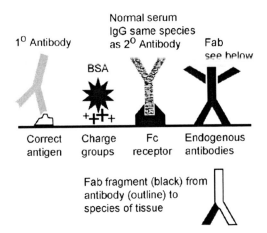

Fig. 5.3 Blocking nonspecific antibody binding. There are blocking agents that will block each of the sites that cause nonspecific binding. The 1° antibody will bind to the correct antigen and is not affected by any of the blocking agents. Charged groups can be quenched by any protein, and BSA is commonly used because it is not a common source of background antibody binding. The Fc receptors must be quenched by the Fc end of an IgG antibody that has no ability to bind other antibodies. Normal serum from the same species as the 2° antibody is commonly used. Endogenous antibodies are blocked by incubations with anti-species Fab fragments that are used only when tissue from injured animals is processed

need to contain goat serum. The blocking solution in this example should contain 1% BSA fraction V and 10% normal goat serum.

The Fc receptors sites are blocked with IgG molecules from any species of animal. The agent most commonly used is 10% *normal serum* (Fig. 5.3). This normal serum or nonimmune serum contains IgGs that will block the Fc receptors. Multiple primary antibodies are used in immunocytochemical experiments and it is possible that the species of the blocking normal serum could create problems. The rule for blocking serum is to use the normal serum from the same animal species of the secondary antibody. For blocking, normal serum must be clear (i.e., not cloudy and not with fibrin strings) and with no pink or red from lysed red blood cells. Cloudy serum indicates precipitated protein that can bind to tissue nonspecifically. Also, there are some reports of "normal serum" available commercially that have antibodies that react in tissue. If background problems persist, one effective strategy is to try different lots of normal serum.

The endogenous antibody sites are blocked with an antibody made against the IgG of the species of the animal. To remove the potential problems of having the species determinant (Fc portion) of the blocking antibody react with other antibodies, the antibody is digested and only the Fab end is used (Fig. 5.3; black Fab fragment). Thus, to block endogenous antibodies in mouse tissue, an anti-mouse IgG *Fab fragment generated in any species* is incubated with the mouse tissue as part of the blocking procedure.

There are additional blocking agents that are sometimes used, but they are needed only for very specific situations. *Glycine* is added because it can bind to free aldehyde groups that occur after formalin or glutaraldehyde fixation. *Gelatin* is sometimes used in place of BSA to bind to charged groups. *Triton X-100* is added to all incubation and rinse solutions help to reduce nonspecific binding. Detergents not only permeabilize cell membranes, but also can act as a blocking agent to lower the binding of some antibodies, probably by reducing nonionic binding (hydrophobic and H-bonding). Finally, *nonfat, freeze-dried milk* (sometimes called "bloto") is used in western or immunoblots but is not recommended for immunocytochemistry. The milk is too strong a blocking agent and actually can inhibit the binding of specific antibodies in the incubation solutions.

Permeabilize Tissue and Cells to Allow Antibody Penetration

For antibodies to penetrate inside fixed cells, the membranes must be opened with detergents. Membranes are *lipid bilayers* that have a hydrophilic or water-soluble side facing the cytoplasm and the extracellular space (Fig. 5.4a). The hydrophobic or water-insoluble sides face each other at the center of the membrane. Also, there are transmembrane proteins that interact with the lipids and are held in the membrane. Membranes are barriers because they do not allow water or hydrated molecules to pass. For immunocytochemistry, the membrane must be partially dissolved to allow antibodies to cross. It is also important that the transmembrane proteins remain cross-linked to other proteins so that they are not washed away (Fig. 5.4b). Detergent will dissolve the membranes but not the transmembrane proteins, which are cross-linked by the fixative to other proteins (e.g., scaffold proteins). For tissue sections, antibodies must penetrate through many cell layers into the center of the section. Achieving this depth of penetration requires removing most of the cell membranes but leaving the proteins so that they can bind antibodies when needed.

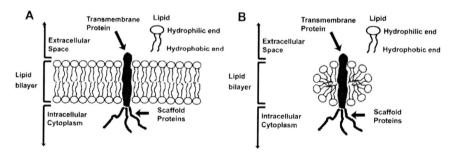

Fig. 5.4 Membrane permeabilization. Membranes are lipid bilayers, with the lipid hydrophobic tail toward other lipid tails and the hydrophilic head toward the cytoplasm or extracellular space. (**a**) Transmembrane proteins pass through the membrane and are held in place by cytoskeletal and scaffold proteins. (**b**) Following fixation, detergent removal of the lipids will leave transmembrane proteins cross-linked to cytoplasmic proteins

Fig. 5.5 Detergents used for immunocytochemistry. (**a**) The nonionic detergent, Triton X-100, has methyl groups that form H-bonds and a hydrophobic tail that solubilizes membranes. (**b**) The ionic detergent, SDS, has a negatively charged group and a hydrophobic tail

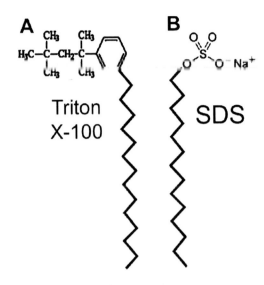

Detergents help dissolve cellular membranes. All detergents are polar lipids that are water soluble with a hydrophilic (polar) ends and can bind the hydrophobic (apolar) ends in the center of the lipid bilayer (Fig. 5.4a). The groups that make up the hydrophilic portion determine the type of detergent. There are two types of detergents, ionic and nonionic (Fig. 5.5).

Nonionic detergents contain methyl groups that participate in hydrogen bonds (Fig. 5.5a) and are able to solubilize membranes but do not destroy protein–protein interactions; examples are Triton X-100 and Tween 20. *Ionic detergents* have highly charged groups (Fig. 5.5b) and are very good solubilizing agents. Ionic detergents not only solubilize membranes, but also destroy native three dimensional protein structures; examples are SDS, deoxycholate, and CHAPS. For immunocytochemistry nonionic detergents are best.

Organic solvents like methanol or acetone also can be used to open membranes, but they also act as denaturing fixatives. Solvents are used at −20°C for 10 min, but the use is not recommended. As described in previous sections, denaturing fixatives may cause loss of epitopes in the cells. Other detergent-like agents, such as saponin or digitonin, are used for transient permeabilization of cell cultures, and reversibly insert into and out of plasma membranes next to cholesterol. These agents must be present in all solutions after the fixative and should only be used for electron microscopy immunocytochemistry.

Using detergents in immunocytochemistry requires decisions not only about type and concentration, but also about the incubation temperature. Detergents do not work the same in fixed cells at room temperature (20°C) and on ice (4°C). Temperatures lower than room temperature change the lipid alignment (phase) in the membrane.

Table 5.1 Detergent extraction of membrane lipids

Detergent	GD1a removed at 4°C	GD1a removed at 20°C
Deoxycholic acid	+	+++
CHAPS	++	++
SDS	–	+++
Triton X-100	+	++
Tween 20	–	+
Saponin	–	+

– no effect, + some loss, ++ moderate loss, +++ heavy loss
Modified from Heffer-Lauc et al., J Histochem Cytochem 55: 805, 2007

A paper on the impact of temperature changes on detergent activity (Table 5.1; Heffer-Lauc et al., 2007), examined the affect of removing lipids from membranes by detergents in tissue from animals fixed with 4% paraformaldehyde. The loss of lipid from membranes was indicated by immunocytochemical labeling for the lipid, GD1a. Removal of lipid from tissue was more effective at room temperature than at 4°C. At 4°C, ionic detergents (deoxycholic acid, CHAPS, and SDS) were able to remove some of the lipids while the nonionic detergents (Triton, Tween) were not able to remove many gangliosides (Table 5.1). However, at room temperature, the ionic detergents were very effective and nonionic detergents were not as effective. Thus, using detergents in the cold is not recommended because they remove less membrane and allow less penetration of antibodies than at room temperature. Also, when samples are processed at 4°C, it is difficult to control the extent of lipid removal when tissues are brought to room temperature for solution changes. In summary, to make the permeabilization more consistent from experiment to experiment, perform detergent incubations at room temperature.

Effects of Blocking Agents on Antibody Penetration

One additional effect of blocking is the ability to change the amount of antibody penetration into tissue sections. Observations of antibody labeling in sections that were incubated with PBS buffer containing 5% normal serum and 10 mg/ml of BSA showed strong labeling in appropriate sites (Fig. 5.6a), while sections incubated with the same concentrations of antibodies and detergent, but without any blocking in the buffer, showed much less labeling in appropriate sites (Fig. 5.6c). These results were collected with wide field fluorescence microscopy, where all of the fluorescence in the section both in focus and out of focus is collected in the image. Examining sections in a confocal microscope will show the labeling only in the plane of focus, also called an optical section. To determine whether the labeling seen in the sections incubated with no blocking had the same distribution of label from the surface of the section into the depth of the section, confocal microscopy was used. When confocal images were taken at the exposed cut surface of the tissue section, labeling was seen that was similar in both sections treated with blocking agents (Fig. 5.6b) and

Wide field fluorescence

Confocal

PBS with blocking

PBS no blocking

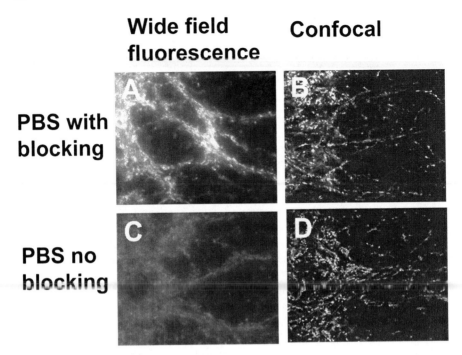

Fig. 5.6 Blocking increases antibody penetration. All micrographs were incubated with the same antibody concentrations and detergent concentrations for the same times. (**a**) Wide field micrograph of section with 5% normal serum and 10 mg/ml BSA. (**b**) Scanning laser confocal micrograph of the same section examined as in a. (**c**) Wide field micrographs of section processed with no blocking serum or BSA and have decreased intensity of labeling. (**d**) Scanning laser confocal micrograph of the same section as **in c, but taken** at the very *top* of the section. Focusing deeper into this section showed the label was only at the surface of the section. Rat spinal cord fixed with 4% paraformaldehyde, 10 μm cryostat section, permeabilized with 0.1% Triton, blocked as indicated, 1° antibody mouse anti-p38 1:1000 overnight, seven rinses, Goat anti-mouse 543 fluorophore 1:1000 for 2 h, six rinses and mount

sections not treated with blocking agents (Fig. 5.6d). When focusing the confocal into the section with no blocking, the label was not observed beyond the surface of the section. Thus, the low level of labeling seen in the wide field fluorescence microscope indicates the total fluorescent labeling in the section and not the intensity of labeling at the surface, where the label was located.

The conclusion from this example is that blocking not only can eliminate nonspecific background, but it can also increase antibody penetration. A mechanism that could explain this observation involves nonspecific charge in the tissue section. In tissue with no blocking the charged groups are not bound or quenched, so it is possible that the charged groups limit its penetration into the tissue. Therefore, an additional way to increase penetration of antibodies into cells and tissues is to use blocking agents.

Combined Incubation Step

The steps of blocking and permeabilization are easily combined into a single step because there is no conflict between the two processes. Perform this step at the beginning of the immunocytochemical procedure just after fixation, rinses, and sectioning. However, it is recommended that detergents be used only in the block/permeabilization step. Adding detergents to antibody incubation solutions can progressively extract more membrane and change the rate of antibody penetration with incubation time. Also, it is possible that some detergents can change final level of antibody binding to epitope, where detergent is present in the antibody incubation solution. To best control for the action of the detergent, perform this as a separate step prior to antibody incubations. Finally, once a specific detergent and incubation conditions are determined, it is important not to change incubation conditions in subsequent experiments without comparing the results to the previous detergent and conditions used.

In practice, including blocking reagents in all incubation and rinse solutions insure that the tissue is adequately blocked. It is recommended that the blocking regents be added to all subsequent incubation steps, and if possible, to the rinse steps between incubations.

Chapter 6
Labels for Antibodies

Keywords Immunohistochemistry · Antibody labeling · Fluorescence microscopy · Fluorescent immunocytochemistry · Fluorescent immunohisto-chemistry · Indirect immunocytochemistry · Immunostaining

Contents

Introduction 55
Fluorescence Theory 56
Four Generations of Fluorescent Labels 58
Immunocytochemistry Fluorophores and Flow Cytometry 59
    Choosing Fluorochromes 61
Enzyme Theory 61
Enzyme Substrates 61
Particulate Label 63
Choice of Fluorescent or Enzymes for Immunocytochemistry 64

Introduction

For years, microscopists have wanted specific stains or labels for different types of cells and different parts of cells. Prior to the mid-1900 s, only dyes were available, many of which stained special categories of compounds. Examples of dyes are hematoxylin that labels nuclei blue and eosin that labels cytoplasmic proteins pink. However, these did not allow labeling of unique proteins or compounds. The first demonstration of a method to label specific proteins came in 1942, when Albert Coons and his colleagues introduced antibody labeling of bacteria within infected human cells (Coons et al., 1942). Coons et al. used an antibody labeled with a fluorescence compound. For the first time, label conjugated to antibodies was used to specifically localize molecules in a microscope. This was the beginning of immunocytochemistry. Today, the strategies for labeling antibodies have continued to evolve and now, many labeling options are available.

R.W. Burry, *Immunocytochemistry*, DOI 10.1007/978-1-4419-1304-3_6,
© Springer Science+Business Media, LLC 2010

Labels are molecules that can be attached to antibodies and will localize the antibodies in cells and tissues. For bright field microscopy, enzymes are the best compounds to label antibodies. Enzymes allow amplification with development of reaction products for labeling. A second approach for bright field microscopy is the use of particulates, such as enhanced gold or silver. The third approach is the popular fluorescent labels. In this chapter, the different types of labels are examined. In subsequent chapters, the methods of applying labeled antibodies will be examined.

Different types of labels or tags attached to antibodies can be divided into three types:

(1) Fluorescence
(2) Enzymes
(3) Particulate label

Fluorescence Theory

Fluorescence is the most common label used to identify antibodies in cells and tissue. *Fluorescent molecules are able to absorb one wavelength of light and emit a higher wavelength of light. Excitation* or absorption is accomplished when a photon of a specific wavelength strikes a fluorophore knocking, an orbital *ground state electron* to that of an *unstable excited singlet electron* (Fig. 6.1a, No. 1). Quickly, the electron loses some energy as heat and falls to a lower state known as the *relaxed singlet electron* (Fig. 6.1a, No. 2). Finally, the *emission* photon occurs when a relaxed singlet electron falls back to its ground state, and its energy is converted into a photon with a lower energy and higher wavelength than the excitation electron (Fig. 6.1a, No. 3). The difference in the wavelengths of the excitation and emission photon is called the *Stokes Shift* (Fig. 6.1b). Remember that the excitation wavelength of a fluorescent compound is of lower wavelength than that of the emission wavelength.

In the fluorescence microscope, photons for excitation and emission can be separated by optical filters. It is useful to look graphically at the excitation and emission spectra of fluorochromes with various light sources. Invitrogen (Molecular Probes) has a web page that is an excellent resource http://probes.invitrogen.com/servlets/spectraviewer?fileid1=11003p72. The example in Fig. 6.2 shows Alexa Fluor 488 excitation in dark dashed line (max 495 nm) and emission in gray dashed line (max 519 nm). The x-axis is the wavelength in nanometers and the y-axis is the intensity. To understand how fluorescence microscopes separate different wavelengths, spectral charts (Fig. 6.2) will be used (Chapter 13, Microscopy and Images).

In selecting fluorescent compounds, their efficiency in conversion of excitation photons to emission photons is important. *The quantum yield of a fluorophore is a measure of the efficiency of conversion from excitation photon to emission photon.* The ratio of the number of photons emitted (emission photons) to the number of

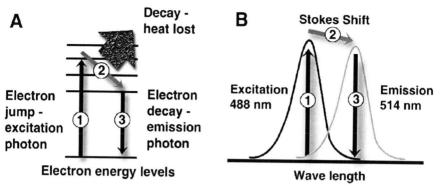

Fig. 6.1 Fluorescent excitation and emission. (**a**) The multiple horizontal lines indicate electron orbitals and an excitation photon, which gives energy to an electron that can jump from a lower orbit to an unstable excited singlet electron (No. 1 in circle). The unstable excited singlet electron loses heat and becomes a relaxed, singlet electron (No. 2 in circle). Finally, the relaxed singlet electron loses its energy as an emission photon and returns to the ground state (No. 3 in circle). (**b**) The difference in wavelengths for the excitation and emission photons shows a shift upward in wavelength that is called the Stokes Shift. The excitation wavelength is always lower than the emission wavelength

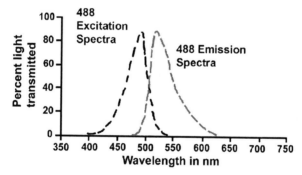

Fig. 6.2 Fluorescent spectra with excitation and emission. The excitation and emission photons are from a range of wavelengths. The percent of photons at each wavelength that can excite a 488 fluorophore are shown in the *black dashed line*. For the emission photons, the percentage of photons emitted at each wavelength is shown. The wavelength of the *top* of the peak is called the maxima

photons absorbed (excitation photons) is the quantum yield, which ranges from a low of 0.05 to a high of almost 1.0. The closer the quantum yield is to 1.0, the better, because the fluorophore will require less excitation for a high level of emission. Quantum yield for both fluorescein and Alexa Fluor is high at 0.7, while rhodamine and the Cy dyes have quantum yields that are low at 0.3. Quantum yield is one of the factors in selection of fluorescent compounds.

Sometimes fluorescence emissions are not observed, even though the correct excitation is used. There are two causes of loss of fluorescence emission. One cause is quenching of emission photons. *Quenching occurs when the energy of the excited singlet is transferred to another very close molecule, with no emission photon from the excited fluorophore.* Quenching, caused by compounds with aromatic rings, oxygen, halides, or proteins is not a common problem with today's immunocytochemistry. Quenching is the basis of the popular technique known as fluorescence resonance energy transfer (FRET), which determines if two proteins tagged with different fluorophores are close enough to transfer the energy.

The other, more common loss of emission photons is called photobleaching. *Photobleaching is the oxidation of the fluorescence molecule so that its structure is changed with its loss of ability to generate emission photons.* The more cycles of excitation and emission, the greater the chance that any single fluorescent molecule could photobleach. Some fluorophores are more susceptible to photobleaching than others. The fluorophore, FITC, photobleaches very rapidly but the fluorophore, Alexa Fluor 488, photobleaches slowly. Photobleaching is the basis of a popular technique to measure movement of proteins in cells. Fluorescent recovery after photobleaching (FRAP) in living cells inactivates fluorescence in a small region of the cell and then tracks the movement of non-photobleached proteins into that region.

Four Generations of Fluorescent Labels

Looking at the fluorescence molecules available today, there is a very wide range of fluorophores. Fluorophores can be classified by year of their introduction. The four generations of fluorescence compounds have increasingly better properties.

1st Generation 1942 – Fluorescein and rhodamine fluorophores. Albert Coons (Coons et al., 1942) prepared the first fluorescent compound used for immunocytochemistry, fluorescein isocyanate. All organic fluorescent compounds use variations of the basic ring structure. The later derived fluorescein isothyocyanate (FITC; Riggs et al., 1958) and rhodamine also photobleach very quickly and are quenched when bound to antibodies.

2nd Generation 1993 – Cy fluorophores; cyanine dyes Cy2, Cy3, Cy5 (Jackson Immuno-Research), AMCA, and Texas Red. These were more photostable (photobleached more slowly than 1st generation) and have high quantum yields.

3rd Generation 1999 – New fluorophores hit the market, Alexa Fluor fluorophores, dyes by Invitrogen/Molecular Probes, ATTO fluorophores by ATTO-Tech, sold by Sigma-Aldrich, and DyLight Fluor sold by Pierce (Thermo Fisher Scientific). This generation has very low photobleaching and is more photostabile. These dyes come in a very wide range of excitation wavelengths from UV to IR, which allow flexibility in selecting fluorochromes to match experimental conditions. Finally, these fluorophore have high quantum efficiencies.

4th Generation 2003 – Quantum dots (Q-dots) heavy metal nanocrystals (Invitrogen/Molecular Probes) fluoresce when exposed to UV light. Quantum dots or Qdots are a cadmium–selenium-rich metallic core with a series of shells that isolate the core and allow attachment of proteins. The size of Qdots ranges between 8 and 30 nm compared with 4 nm for an IgG with Alexa Fluor. Qdots are excited at low wavelengths (350–425 nm) and emit at a high wavelength, which means that new filter sets are required for fluorescence microscopy. A major advantage of Qdots is that they do not photobleach. For immunocytochemistry, Qdots have penetration problems because of their large size and as a result they are only of limited use. When better methods of coating become available to help achieve smaller molecular size, Qdots will become more useful for immunocytochemistry.

Immunocytochemistry Fluorophores and Flow Cytometry

Regarding fluorophores, it is important to say something about flow cytometry and compare it to immunocytochemistry. Many scientists begin their use of fluorescent-labeled antibodies while using flow cytometry and then shift to immunocytochemistry. Flow cytometry is an analytical technique for evaluating labeled cells as they pass rapidly through a beam of light. Many of the fluorophores used in flow cytometry are commonly thought of as making good labels for immunocytochemistry. However, because of fundamental differences between flow cytometry and immunocytochemistry techniques, there are certain fluorophores that are unsuitable for use in immunocytochemistry.

First, in flow cytometry, the individual cells are in the light path for milliseconds. The rapid flow combined with short exposure to light eliminates problems due to photobleaching. Thus, the fluorophore fluorescein, one of the least costly fluorophores available, works well because of the short exposure to light even though it photobleaches very rapidly. In immunocytochemistry, photobleaching is a major problem.

A second issue occurs when several labels are needed in flow cytometry and a single laser must be used to excite these fluorophores. Here, different fluorophores are excited at one wavelength, for example a 488 nm laser, and the fluorophores have emissions at different wavelengths. The fluorophores then are shifted or tandem conjugate where the excitation is based on one fluorophore and the emission is based on a second covalently bound fluorophore. There are several excitation fluorophores that increase in the Stokes Shift for emission to much higher emission wavelengths. To serve as a standard for shifted fluorophores, Alexa Flour 488 is excited at 495 nm and emits at 519 nm (Fig. 6.3). Phycoerythrin (PE) a pigment from blue-green algae is excited at 488 nm and has the emission at 575 nm (Fig. 6.3; Glazer and Stryer, 1983). To give different emission wavelengths, PE can be bound to Texas Red with an emission of 630 nm. PE can be bound to Cy5 with the emission 665 nm, or bound to Cy5.5 with the emission 720 nm (Herzenberg et al., 2002). The advantage is that instruments with one 488 laser can give

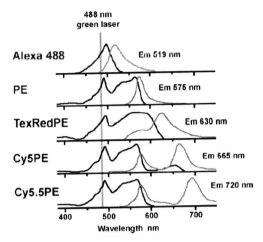

Fig. 6.3 Fluorophores used in flow cytometry. With a single laser line at 488 nm, several fluorophores can be excited (*black solid line*). These fluorophores emit (*gray solid line*) at a different range of wavelengths (Em). Alexa Fluor 488 spectra emission of 519 nm shows the expected Stokes Shift. Phycoerythrin (PE) has a larger Stokes Shift and is used as a basis for other fluorophores with emissions shifted to even higher wavelengths. Texas Red and PE (TexRedPE), Cy5 and PE (Cy5PE), and Sy5.5 and PE (Cy5.5PE) combinations demonstrate increased shift of emissions to higher wavelength. These fluorophores are not useful for immunocytochemistry

emissions at multiple different wavelengths (here PE alone and three shifted fluorophores), allowing multiple labels to be detected. The problem with these shifted or tandem conjugated fluorophores is that if multiple lasers are used then these shifted fluorophores are capable of generating bleed-through and detecting multiple labels in one channel. This is the problem for confocal microscopy and wide field fluorescence microscopy.

There are other excitation compounds that can be bound to different fluorophores with different emission wavelengths. Allophycocyanin (APC) is excited at 633 nm and emits at 647 nm. Binding APP to Cy5.5 retains the 633 nm excitation but generates an emission at 705 nm; binding APC to Cy7 generates an emission at 750 nm. A third compound is PKH, named for its founder Paul K. Horan; PKH is a membrane dye that intercalates into the hydrophobic region of membranes. PKH2 and PKH67 are excited at 490 nm with emission at 504 nm, while PKH26 is excited at 551 nm and emits at 567 nm. These dyes have excitation spectra that are very broad, over 100 nm, making them difficult to use with multiple fluorescent labels for immunocytochemistry.

Before using a fluorophore-labeled antibody from flow cytometry for immunocytochemistry, check the spectra of fluorescent labels to rule out the possibility of bleed-through. Remember that although flow cytometry and immunocytochemistry use fluorescent dyes, the characteristics of these dyes are optimized for the technique and might not translate well when moved from flow cytometry to immunocytochemistry.

Choosing Fluorochromes

The following guidelines will help in choosing fluorophores.

- Match the correct microscopic filter set with excitation and emission wavelengths of the fluorochromes selected.
- Use third-generation fluorochromes for longest and brightest label.
- For cells having transfected green fluorescent proteins (GFPs) variants omit fluorophores with excitation wavelengths in the same range as the GFP variant used.
- For multiple fluorochromes, a standard series of emissions with separation includes 350 nm (blue), 488 nm (green), 546 nm (red), and 633 nm (high red).
- Avoid Texas Red (491 nm) because fluorophores excited by either 546 or 633 nm cannot be used in the same experiment due to excitation–emission overlap.
- Fluorophores used for flow cytometry might not be appropriate for immunocytochemistry.

Enzyme Theory

The advantage of enzymes as labels for immunocytochemistry is signal amplification, which generates large amounts of reaction product from one enzyme. When an enzyme is attached to an antibody, incubations with labeled antibodies are similar to incubations with fluorescence-labeled antibodies, except that an additional step is required that will incubate the enzyme with clear substrate to allow production of a colored reaction product. A single enzyme can catalyze the generation of many thousands of reaction products, thus increasing the sensitivity of detection. The best use of enzymes is for detection of low-copy antigens when high level of sensitivity is needed. Enzymes markers are not visible by themselves, but require a development step with one of several methods with the resulting label imaged in a bright filed microscope and not a fluorescence microscope.

Enzyme Substrates

The enzyme most commonly used is *horseradish peroxidase* (*HRP*), which is isolated from the root of the horseradish plant. HRP is a 40 kDa enzyme that is bound to antibodies with no loss of either enzyme activity or antibody binding. HRP enzymatic activity is an oxidation reaction that is used for a wide variety of reactions where oxidation gives rise to labeling.

The most important class of substrates for HRP is *chromogens* that generate an optically dense reaction product for bright field microscopy (Nakane and Pierce 1966). The chromogen $3'$-$3'$ diaminobenzidine (DAB; 224 kDa) is the most commonly used substrate and gives an optically brown reaction product. The colorless DAB and hydrogen peroxide (H_2O_2) bind to HRP and the DAB is oxidized, yielding a brown reaction product (Fig. 6.4). HRP can generate reaction product for

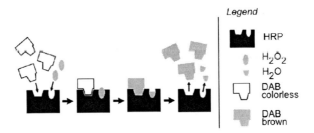

Fig. 6.4 Enzyme labeling. Horseradish peroxidase (HRP) is attached to an antibody and serves as an enzyme label. HRP changes a colorless chromogen (DAB) into a brown substance seen in the microscope. The catalyst for the HRP reaction is hydrogen peroxide (H_2O_2) that is converted to water (H_2O)

15 min, and the buildup of brown reaction product is extensive. Warning: DAB is carcinogenic, follow protocols for handling carcinogenic compounds.

There are several chromogens that can generate reaction products, each with slightly different characteristics. One substrates is tetramethylbenidine, **TMB** (3,3′,5,5′-tetramethylbenzidine), which generates a water-soluble blue product that must be precipitated by dehydration after enzyme reaction. Because of the increased detection sensitivity, the incubation with TMB might give high background. In these cases, the antibody can be used at higher dilutions, saving on use of the antibody. Also, TMB is more quickly oxidized than other HRP substrates, resulting in faster color development.

There are intensification methods for HRP reaction products that involve heavy metals. DAB–nickel enhancement or nickel–DAB (*NiDAB*) gives a darker, more stable reaction product signal (Angelov et al., 1998). When nickel (II) sulfate is used with DAB, it requires a buffer system that contains no chloride ions, since they precipitate nickel. For other HRP substrates, the reaction products are best retained if the sections are dehydrated and mounted in resin-based organic medium, such as Permount. The advantages of NiDAB are that the chromogen color is blue-black and the spread of the reaction product is less than DAB alone. Alternate substrates are DAB–Co (0.5% CuSO$_4$) and DAB–imidazole, which increase detection intensity.

A last method of localizing HRP is by oxidizing *ionic silver* to give a visible deposition of elemental silver as a label. This method uses buffers and reagents that require a low pH and cause destruction of cellular morphology. Therefore, this method is not recommended except for specific cases where discrete particulate label is needed.

In some cases, chromogen development generates significant background, which can be reduced by use of *glucose oxidase*. Glucose oxidase catalyzes the conversion of substrate β-D-glucose to D-gluconic acid with the generation of H_2O_2, which in turn is used by HRP to catalyze the DAB reaction. Glucose oxidase can be used as a soluble reagent with the DAB-labeled antibodies to label cells (Shu et al., 1988). If the glucose oxidase is attached to a 2° antibody the resolution is improved (Kuhmann and Peschke, 1986).

One problem with enzyme labeling is that the dense reaction product can diffuse within cells, masking the precise location of the antigen. Diffusion of the reaction products might limit resolution and obscure the underlying cellular structure. For restricted distribution of the reaction product use NiDAB (Angelov et al., 1998).

Enzyme amplification methods increase the detection sensitivity but not antibody sensitivity (detection of antigens by antibodies) of the immunocytochemistry method. While the labeling is increased with enzyme reaction product, the amount of antigen for the primary antibody does not change even though more label is seen that with non-amplified systems. This is important because enzymes can generate visible reaction with very little antigen bound by the primary antibody. Also, it is important to distinguish the sensitivity of the detection system from that of the number of antibodies bound to the tissue. Increasing the detection sensitivity will only show the location of labeled antibodies. The binding of antibodies to antigens in tissues is regulated by the number of antigens in the tissue and the dilution of the antibody. A large amount of label does not indicate large amount of antigen.

Multiple 1° antibodies with enzymes present unique problems. While the methods of multiple 1° antibodies will be discussed in Chapter 11, Multiple 1° Antibodies Different Species, there are unique challenges using multiple 1° antibodies with enzyme labels. Because the HRP reaction product can cover the protein bound by the 1° antibody, the dense reaction product can also prevent binding of a second 1° antibody located within the reaction product. The first chromogen development of a multiple 1° antibody experiment will cover the area labeled with a sticky-dense reaction product that prevents other antibodies from binding antigens in the labeled area. Thus, if colocalization of two antigens is needed, it might not be detected with an enzyme-based detection system.

With HRP, multiple 1° antibodies can be detected following developing of each antibody with a different colored chromogen. Multiple chromogens involved in colocalization can complicate the user's ability to distinguish the colors with bright field microscopy. Colocalization requires the ability to determine the amount of each color that has contributed to the label for a specific structure. Recently, a spectral camera has been used to successfully separate different color chromogens developed from HRP (van der Loos, 2008). The ability to separate different fluorophores is much easier because the fluorescence microscope is designed to distinguish different wavelengths and the bright field cameras are not.

Finally, for multiple 1° antibodies, the first HRP enzyme must be inactivated, so that it will not participate in development when the second 1° and 2° with HRP is used. One way to eliminate this problem is to use a different enzyme, alkaline phosphatase, which has different substrates from HRP. A discussion of endogenous peroxidase activity is found in Chapter 8, Controls.

Particulate Label

A third class of labels for antibodies that can be used for bright filed microscopic immunocytochemistry is particulate labels. Antibodies can be labeled with gold particles (nanogold), which are too small to be seen without electron microscopy. However, several silver enhancement methods can develop the gold into larger particles that are seen in the light microscope. These methods greatly increase the detection efficiency and make it possible to detect single antibodies, but should be reserved for times where detection is the major problem because there are many difficulties unique to the silver enhancement of gold. The method will not be discussed in this book.

Choice of Fluorescent or Enzymes for Immunocytochemistry

Immunocytochemistry labels provide a dizzying list of options. Below are the most important advantages (+) and disadvantages (−) of these two methods.

Enzyme methods with chromogen:

+ Ideal for low magnification and identification of whole cells.
+ Reaction product will not deteriorate over time.
+ High detection sensitivity will detect low-copy proteins.
+ Counter staining with dyes shows unlabeled cells.
+ Can be viewed in a bright field microscope that is commonly available.
− Low detection resolution hampers the localization of proteins.
− Overlapping different colors for multiple primary antibodies makes them difficult to distinguish.
− Dense, sticky reaction product inhibits colocalization with multiple primary antibodies.
− Not visible in confocal microscopes.
− Requires enzyme development, which has its own problems.
− Many chromogens have health hazards.

Fluorescent method:

+ Ideal for high magnification because of its excellent detection resolution for protein labeling.
+ Can be viewed in a scanning laser confocal microscope.
+ Compatible with observing GFP-labeled proteins.
+ Amplification methods provide high sensitivity.
+ Can be used to distinguish multiple primary antibodies (easily up to four) with different labels.
− Not easily seen label at low magnification.
− Fluorescence is not permanent.
− Requires special filters to examine with a fluorescence microscope.
− Multiple primary antibody detection can have bleed-through of labels.
− Counterstain of unlabeled cells is limited to single compartments, like nuclei.

Chapter 7
Application Methods

Keywords Immunohistochemistry · Antibody labeling · Fluorescence microscopy · Fluorescent immunocytochemistry · Fluorescent immunohisto-chemistry · Indirect immunocytochemistry · Immunostaining

Contents

Introduction 66
Direct Immunocytochemistry 66
   Direct Immunocytochemistry Advantages 67
   Direct Immunocytochemistry Disadvantages 67
Indirect Immunocytochemistry 67
   Indirect Immunocytochemistry Advantages 68
   Indirect Immunocytochemistry Disadvantages 68
Avidin–Biotin Molecules 68
Direct Avidin–Biotin Immunocytochemistry 69
   Direct Avidin–Biotin Method Advantages 70
   Direct Avidin–Biotin Method Disadvantages 70
Indirect Avidin–Biotin Immunocytochemistry 70
   Indirect Avidin–Biotin Advantages 71
   Indirect Avidin–Biotin Disadvantages 71
Avidin–Biotin Complex (ABC) Immunocytochemistry 71
   Avidin–Biotin Complex (ABC) Advantages 73
   Avidin–Biotin Complex (ABC) Disadvantages 73
Tyramide Signal Amplification (TSA) Immunocytochemistry 73
   Tyramide Signal Amplification Advantages 74
   Tyramide Signal Amplification Disadvantages 75
ABC with TSA 75
   ABC with TSA Advantages 77
   ABC with TSA Disadvantages 77

R.W. Burry, *Immunocytochemistry*, DOI 10.1007/978-1-4419-1304-3_7,
© Springer Science+Business Media, LLC 2010

Introduction

There many different ways to use labels or tags attached to antibodies for immunocytochemistry. Initially, in 1942, the label was attached directly to the antibody,
which bound to the antigen. Since then, many different methods have been developed to indirectly bind labeled antibodies to the antigen. More recent methods have
increased sensitivity in detecting low frequency antigens. Remember, the antibody
that binds to an antigen is called the *primary* (1°) *antibody* and it has the specificity for the antigen. Location of the 1° antibody is determined by one of several
application methods. It is also worthwhile to remember that labeled antibodies have
multiple label molecules attached, and although the antibody drawing in this book
shows only one label is conjugated (i.e., attached) to each antibody.

Direct Immunocytochemistry

For direct immunocytochemistry, the label is bound to the 1° *antibody.* For example, to locate an antigen in cells, the 1° antibody would be a rabbit anti-antigen
labeled with fluorophore (Fig. 7.1). Such a procedure with directly conjugated antibodies requires only the 1° antibody and no additional antibodies. Thus, the direct
immunocytochemical procedure is the simplest method, and historically, was the
first method for immunocytochemistry.

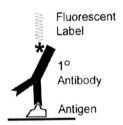

Fig. 7.1 Direct
immunocytochemistry. In the
direct immunocytochemistry
method, the 1° antibody binds
its antigen and the label is
directly attached to it

For direct immunocytochemistry, conjugating markers to 1° antibodies requires
large amounts of purified antibody for the labeling reaction. It is rarely worth the
effort to label each primary antibody because of the large number of 1° antibodies
and the work need to label and purify each antibody. Today, direct immunocytochemistry is not the method of choice, but some vendors sell a lot of labeled 1°
antibodies for common antigens. The source of the labeled direct antibody is a large
animal (e.g., goat, sheep) or a monoclonal antibody, where large amounts starting material can be obtained. The directly conjugated antibodies are used in flow
cytometry and they can be used for immunocytochemistry.

Direct Immunocytochemistry Advantages

- Less complex experiments with fewer steps required.
- For multiple 1° antibody experiments, labeled 1° antibodies from the same species can be used with no problems.

Direct Immunocytochemistry Disadvantages

- Production of labeled 1° antibodies is more time consuming and expensive than unlabeled 1° antibodies.
- Labeled 1° antibodies give no flexibility with selecting different labels to fit multiple 1° antibody experiments.
- Weaker label detection results because only label from the 1° antibody with no amplification from additional label bind to the 1° antibody.

Indirect Immunocytochemistry

For indirect immunocytochemistry, the labeling comes from a labeled secondary (2°) antibody that binds the 1° antibody in a second incubation step (Fig. 7.2). The 2° antibody is generated by injecting purified IgG from one species of animals as the antigen. These *2° antibodies bind to the Fc end of the 1° antibody, which is common to all IgG molecules from a single species.* Thus, a 2° antibody is made against all IgG molecules from one species of animals, and is therefore species specific. For example, in an indirect immunocytochemical procedure, the antigen is bound by a 1° antibody, rabbit anti-antigen (unlabeled), and the 2° antibody is a goat anti-rabbit labeled with 488 fluorophore (Fig. 7.2). Although not shown here, multiple 2° antibodies will bind to a single 1° giving amplification of the signal as compared to direct immunocytochemistry. The 2° antibodies are made in large animals like goat or donkey and are economical to produce.

Fig. 7.2 Indirect immunocytochemistry. An unlabeled 1° antibody binds to the antigen and a labeled 2° antibody binds to the 1° antibody. The labeled 2° antibody is made in a different species of animals against the IgG species of the 1° antibody. A variety of labels can be used from fluorescence to enzymes

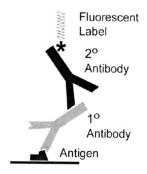

Fluorescent Label

2° Antibody

1° Antibody

Antigen

A major advantage to the indirect method is that each labeled 2° will attach to all 1° antibodies from one species (e.g., goat anti-rabbit IgG labeled with 488 fluorophore works for all rabbit 1° antibodies). This is the method of choice today because so many different antibodies are used in biomedical research. Indirect immunocytochemistry detects proteins in cells with the high detection sensitivity, good flexibility of reagents, and the fewest steps.

Indirect Immunocytochemistry Advantages

- Labeled 2° antibodies are an inexpensive way to locate 1° antibodies.
- Commercial 2° antibodies are available with many possible labels.
- Species-specific 2° antibodies increase the flexibility in labeling different 1° antibodies.
- Increased labeling over direct immunocytochemistry occurs because several 2° antibodies can bind a single 1° antibody.

Indirect Immunocytochemistry Disadvantages

- Additional controls are needed to show that the 2° antibodies are binding to the correct 1° antibody
- Multiple 1° antibodies used together in one experiment must be generated in different species of animals.

Avidin–Biotin Molecules

Use of avidin–biotin molecules in immunocytochemistry dramatically increases the detection sensitivity and greatly increases the flexibility of labels. The power of avidin–biotin is their ability to bind with extremely high affinity to each other. In addition, the ligand (biotin) and receptor (avidin) are unique for each other. There are multiple ways that labels can be conjugated to avidin or biotin and then used in immunocytochemistry. This section will introduce these molecules and show a several methods for their use in immunocytochemistry.

The power of *avidin is in its four high-affinity binding sites for biotin* (Fig. 7.3). Avidin is a protein from egg white (68 kDa). A similar protein is streptavidin from *Streptomyces avidinii* bacteria (75 kDa). Avidin can show nonspecific binding to cells because it has charged sugars attached. Streptavidin, however, is not glycosylated, and therefore is the preferred reagent because it has lower nonspecific binding. Finally, it is easy to attach to proteins and antibodies covalently to avidin without changing its ability to bind biotin.

Biotin is a very small (244 Da) molecule, also known as vitamin B7, that binds avidin with extremely high affinity. Each biotin has just one avidin-binding domain

Fig. 7.3 Avidin–biotin reagents. Avidin is a protein from either egg whites or from bacteria and is not found in tissues or cells. Biotin, a very small molecule, is also known as vitamin B_7. Four biotin molecules can attach to one avidin, with an affinity that is extremely high making the binding almost permanent

Table 7.1 Properties of avidin and biotin

	Avidin	Biotin
Mol wt.	68,000–75,000 Da	244 Da
Binding sites	Four for biotin	One for avidin
Source	Egg white, bacteria	Mammalian vitamin
Covalent binding to antibodies	Yes	Yes

(Fig. 7.3). Avidin binds biotin at one end and at the other end has a linker for covalent conjugation to antibodies.

In immunocytochemistry avidin–biotin binding, while not covalent, is essentially permanent. Also, since avidin has multiple binding sites for biotin, this can amplify the labeling.

Direct Avidin–Biotin Immunocytochemistry

Direct avidin–biotin method is a three-step procedure, using biotin-labeled 2° antibody to bind the 1° antibody (Fig. 7.4). In a third incubation step, an avidin bound to a label binds to the biotin of the 2° antibody, supplying the label to locate the 1° antibody. One advantage of this method is the flexibility with labels for a single 2° antibody. Because the 2° antibody does not contain the label, the number of 2° antibodies needed is small. Rather than purchasing 2° antibodies against each species with four different labels, only one antibody with biotin is needed. Also, purchasing avidin with four different labels allows use with any species of 2° antibody, further reducing the number of 2° antibodies purchased. Thus, the flexibility comes from being able to mix-and-match combinations of biotin-labeled 2° and labeled avidin. For example, rather than having a 488 fluorophore-labeled 2° for each species of 1° (e.g., mouse, rabbit, goat), only an avidin-labeled 488 fluorophore is needed for all species 2° antibodies.

Fig. 7.4 Direct avidin–biotin immunocytochemistry. An unlabeled 1° antibody binds to the antigen and a 2° antibody labeled with biotin binds to the 1° antibody. Next, a labeled avidin binds to the biotin. This method has three incubation steps

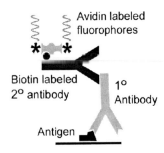

Direct Avidin–Biotin Method Advantages

- Increases flexibility by using a series of labeled avidin molecules that are the same for all species of 2° antibodies reducing costs.
- Offers modest increase in sensitivity over indirect immunocytochemistry with fluorescent-labeled 2°.

Direct Avidin–Biotin Method Disadvantages

- Requires additional reagents.
- Requires an additional incubation step over indirect immunocytochemistry with fluorescent-labeled 2°.
- Requires additional controls.

Indirect Avidin–Biotin Immunocytochemistry

Indirect avidin–biotin method uses two additional incubation steps after the 2° antibody (Fig. 7.5). Avidin with no label is incubated after the 2° antibody. In fourth step, a labeled biotin is incubated with the complex and it binds to all unbound sites on the avidin. The avidin then serves as a bridge to attach the 2° antibody to the fluorophore with sequential incubations of these reagents.

One advantage of indirect avidin–biotin method is an amplification of the detection system. For example, after biotin conjugated 2° antibody, avidin is incubated, binding to the available biotins. After the remaining free avidin is rinsed off, a biotin conjugated to 488 fluorophore is added and it binds to the remaining sites on the avidin. With this method, amplification occurs because any avidin can be attached to multiple biotins conjugated with either fluorescent or enzyme. This method requires two additional incubation steps of the indirect immunocytochemistry with fluorescent-labeled 2°.

Another advantage is flexibility in reagents. Because the biotin is labeled with the fluorophore, both the avidin and the 2° antibody need not be labeled. Rather than

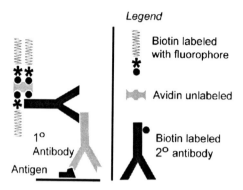

Legend

Biotin labeled
with fluorophore

Avidin unlabeled

Biotin labeled
2° antibody

1°
Antibody

Antigen

Fig. 7.5 Indirect avidin–biotin immunocytochemistry. Following the 1° antibody binding to the antigen and a 2° antibody labeled with biotin binds to the 1° antibody. Unlabeled avidin is incubated in a third step and binds the biotin-labeled 2° antibody with several biotin-binding sites still unoccupied. In a fourth step, labeled biotin is bound to the avidin

purchasing as complete set of labeled 2° antibodies with each fluorescent label, the only 2° antibody needed is biotin labeled.

Indirect Avidin–Biotin Advantages

- Biotin conjugated to labels (fluorophores or enzymes) with the unlabeled avidin bridge amplifies the label.
- Labeled biotin gives flexibility of having a set of labeled biotin that can be used for any species 2° antibodies giving the greatest flexibility for combinations of labels of any method.

Indirect Avidin–Biotin Disadvantages

- Requires two additional incubation steps over indirect immunocytochemistry.
- Additional controls are needed.

Avidin–Biotin Complex (ABC) Immunocytochemistry

Avidin–biotin complex (ABC) uses a reagent made from avidin, biotin, and HRP for a much larger increase in detection sensitivity (Hsu et al., 1981). ABC is based on the molecular complex that is made by mixing HRP-bound biotin with an excess of unlabeled avidin (Fig. 7.6). The proportions of biotin HRP and avidin are critical for developing the complex size that is needed for immunocytochemistry. The vendor, Vector Laboratories, produces an ABC reagent that is easy to use and consistent.

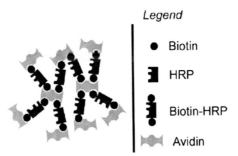

Fig. 7.6 Avidin–biotin complex (ABC). To increase the number of HRP enzymes bound to a 1° antibody an avidin–biotin complex (ABC) is used. To make the complex, first multiple biotins are bound to HRP and then this is incubated with a dilute avidin. All of the HRP–biotin reagents bind to avidin, generating complexes with unbound avidin still available. This reagent is now used in ABC immunocytochemistry

The ABC contains numerous HRP molecules for label generation, and unbound avidin available to bind a biotin 2° antibody.

ABC is a very popular way of amplifying the HRP label for light microscopy. For example, an experiment uses a rabbit anti-antigen 1° antibody and a 2° goat anti-rabbit antibody labeled with biotin (Fig. 7.7). When the ABC complex is added, free avidin on the surface of the complex binds to biotin on the antibody, which carries many more HRP molecules than a single HRP-labeled 2° antibody. Vector Laboratories makes ABC kits that use dropper bottles to measure reagents. Such kits are almost foolproof.

While the use of the ABC increases detection sensitivity by adding HRP enzymes (Fig. 7.7) to the labeling process, it does not change either the number of the 1°

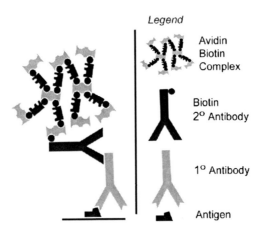

Fig. 7.7 Avidin–biotin complex (ABC) immunocytochemistry. After the 1° antibody binds the antigen and the 2° antibody labeled with biotin binds to the 1° antibody. The third incubation is with the ABC reagent that labels with an enzyme followed by chromagen development

antibodies binding to the antigen or the antibody sensitivity. This point is important because increased labeling with ABC does not mean that more antigens are being detected. The sensitivity of the 1° antibody is the same regardless of the detection method used. The increase in sensitivity seen with ABC simply enhances the ability to locate the 1° antibody.

Avidin–Biotin Complex (ABC) Advantages

- ABC greatly improves detection sensitivity over indirect immunocytochemistry.
- Levels of 1° antibodies use for incubation can be decreased by 10- to 100-fold because of increased detection.

Avidin–Biotin Complex (ABC) Disadvantages

- Requires additional incubation steps over indirect immunocytochemistry.
- Label levels are not reliable indicators of the amount of antigen present.
- Additional controls are needed.

Tyramide Signal Amplification (TSA) Immunocytochemistry

Tyramide signal amplification (TSA) is a powerful amplification method that allows fluorescent labeling to be significantly amplified. TSA is a patented technology from Perkin Elmer that has been licensed to other companies for use in reagents.

The key to this detection method is the very small compound, tyramine, with two reactive groups (Fig. 7.8). Tyramine contains an amine (-NH$_2$ group), which binds a label (e.g., fluorophore or biotin) on one end and contains a hydroxyl (-OH group) on the other end. When bound to a label via the amine group, the molecule is called tyramide. In the presence of HRP, the inactive tyramide has the OH group, which now binds to a 1° amine on any protein of the tissue. For example, the 1° antibody, rabbit anti-antigen is bound by the 2° antibody, goat anti-rabbit

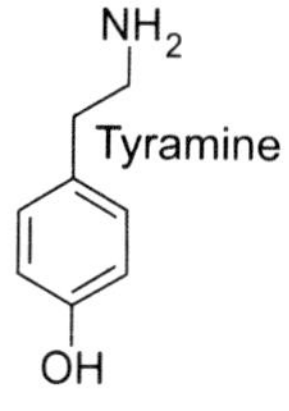

Fig. 7.8 Tyramine. The chemical tyramine has an amine group (-NH$_2$) used to bind to a label with the reagent now called tyramide. The hydroxyl group (-OH) remains unbound for use in immunocytochemistry

Fig. 7.9 Tyramide signal amplification (TSA) immunocytochemistry. The 1° antibody binds to the antigen and a 2° antibody labeled with HRP binds to the 1° antibody. The labeled tyramide reagent is added with a small amount of the catalyst H_2O_2. The enzyme HRP converts the inactive labeled tyramide into a reactive labeled tyramide that binds to the amino acid tyrosine found in proteins

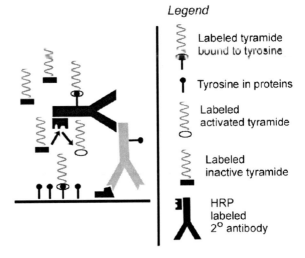

Legend

Labeled tyramide bound to tyrosine

Tyrosine in proteins

Labeled activated tyramide

Labeled inactive tyramide

HRP labeled 2° antibody

labeled with HRP (Fig. 7.9). Next, is incubation with tyramide bound to 488 fluorophore in the presence of H_2O_2 causes the HRP to enzymatically activate tyramide (Fig. 7.9). The tyramide changes the -OH group to a highly reactive group with a brief life time. This activated tyramide can for a very brief time bind to any amine most commonly the tyrosine amino acids. Because there are numerous tyrosine amino acids surrounding the antigen, all activated tyramide–fluorochrome molecules will bind near the antibody-bound antigen. Other activated tyramide molecules will bind to each other but not to tissue and later, these will be washed out with rinses.

There are three attributes of the TSA procedure that makes the method very powerful. First, the activated tyramide reacts quickly with tyrosine and cannot diffuse far from the HRP-labeled 2° antibody. Second, is that the tyramide is bound to the tissue and is not free to move after the reaction is completed. These advantages are compared to development of many chromogens by HRP where the reaction product is not bound to the tissue and is able to move. Third, there are a wide variety of label compounds that can be attached to tyramine, including both fluorophores and biotin. It is possible to use tyramide bound to biotin and then an avidin HRP, so that a chromogen can be used to localize the 1° antibody.

Tyramide Signal Amplification Advantages

- Increased detection sensitivity provides the ability to detect infrequent antigens or antibodies at high dilution of the 1° antibody.
- TSA is the best amplification method for fluorescent immunocytochemistry.
- There is a small spread of fluorescent label from the 1° antibody.

Tyramide Signal Amplification Disadvantages

- TSA does spread the label from the location of the 1° antibody decreasing label resolution.
- Label levels are not reliable indicators of the amount of antigen present.
- Background label is also amplified.
- Additional controls are needed for TSA.

ABC with TSA

ABC with TSA offers the largest amount of label amplification and therefore the best detection sensitivity. Combining the two high-detection-sensitivity methods into one procedure when the antigen is present in the tissue at low number of copies or when there are very high dilutions of 1° antibodies (e.g., 1:500,000). In both cases, the sensitivity of the detection must be maximized.

The procedure begins with the standard ABC protocol, where biotin-labeled 2° antibody is bound with the avidin–biotin complex (ABC). This adds a significant number of HRP molecules for each 1° antibody. To start the TSA method, incubate tissue with biotin-labeled tyramide with H_2O_2; the HRP enzymes generate activated tyramide (Fig. 7.10). The activated tyramide then binds to the available tyrosine amino acids on cellular proteins near the antigen, on the antibodies, and even on

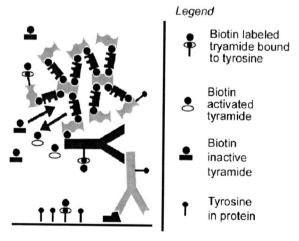

Legend

- Biotin labeled tryamide bound to tyrosine
- Biotin activated tyramide
- Biotin inactive tyramide
- Tyrosine in protein

Fig. 7.10 ABC and TSA. To obtain the largest amount of signal for a single 1° antibody combine ABC and TSA methods. After the 1° antibody binds the antigen and the 2° antibody labeled with biotin binds to the 1° antibody. The third incubation is with the ABC reagent that labels with an HRP. In a fourth incubation, HRP activation of biotin-labeled tyramide near the antigen, adds numerous biotins to the tissue near the antigen. The labeling is completed with fluorescence (Fig. 7.11) or enzymes (Fig. 7.12)

Fig. 7.11 ABC with TSA detection with fluorescence. With the biotin tyramide generated in the ABC and TSA (Fig. 7.10), detection is with an avidin with fluorescent label. The fifth incubation will give the fluorescent labeling

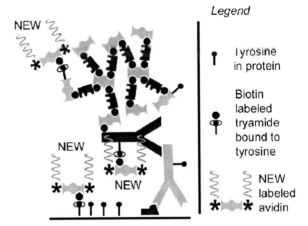

Legend

↑ Tyrosine
  in protein

Biotin
labeled
tryamide
bound to
tyrosine

NEW
labeled
* avidin

the avidin (Fig. 7.10). The result is that numerous biotin molecules are bound to the tissue around the 1° antibody.

The next step is to add avidin bound to fluorescent label (Fig. 7.11) or HRP label (Fig. 7.12). The 488 fluorophore avidin will bind to the biotin bound to tyrosine (Fig. 7.11), giving a large increase in labeling over standard indirect immunocytochemistry. The HRP avidin further increases the number of HRP molecules surrounding the antigen (Fig. 7.12). When the chromogen development is then performed, a very large amount of reaction product will be deposited and the location of one 1° antibody is likely to be detected.

Two concerns with the ABC with TSA method are related to the ultra high dilutions of 1° antibody and the incomplete labeling of all the antigens within the cell.

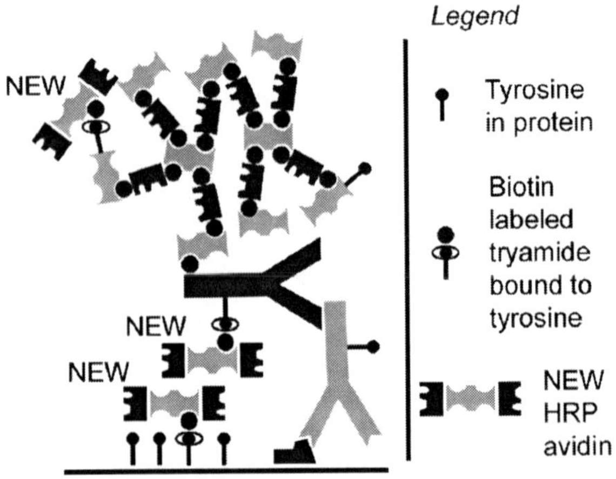

Legend

↑ Tyrosine
  in protein

Biotin
labeled
tryamide
bound to
tyrosine

NEW
HRP
avidin

Fig. 7.12 ABC with TSA detection with HRP. With the biotin tyramide generated in the ABC and TSA (Fig. 7.10) detection is with an avidin-labeled HRP. Following the fifth incubation with avidin–HRP, the HRP is developed to generate a visible chromogen

With 1° antibody dilutions so high (e.g., 1:1,000,000), it is possible that the 1° antibody is not saturating the possible antigens. The antigens not bound will give false negative results. One solution to this problem is to show the same labeling with both high and low dilutions and thus the 1° antibody at high dilution is not limiting the ability of the detection system to localize antigens.

ABC with TSA Advantages

- Provides dramatic increases in the detection sensitivity.
- Provides the ability to detect very rare antigens
- Allows use of extremely highly diluted 1° antibodies.

ABC with TSA Disadvantages

- Label levels are not reliable indicators of the amount of antigen present.
- Discrete labeling of small structures is not possible.
- Additional controls are needed.

Chapter 8
Controls

Keywords Immunohistochemistry · Antibody labeling · Fluorescence micro-scopy · Fluorescent immunocytochemistry · Fluorescent immunohistochem-istry · Indirect immunocytochemistry · Immunostaining

Contents

Introduction 79
Three Immunocytochemistry Controls 79
    1. 1° Antibody Controls 80
    2. 2° Antibody Controls 84
    3. Labeling Controls 85

Introduction

Here is the fundamental issue: Can the labeling in immunocytochemistry images be trusted? Does the antibody location reveal the antigen being investigated? It is easy to think that a micrograph is naturally correct, because how would the label end up in the wrong place? Scientists experienced with immunocytochemistry accept the validity of localization, but only when appropriate controls are performed. Too often, initial excitement from an apparent positive result is dashed when control experiments show the labeling was not specific. Control experiments are essential to rule out the possibility of a nonspecific labeling being identified as specific labeling. In this chapter, three types of control protocols are presented. Together, these protocols are necessary to validate the specificity of labeling and eliminate the consideration of labeling as a misrepresentation of normal localization.

Three Immunocytochemistry Controls

The published literature has a variety of different names for immunocytochem istry controls. Some of these controls are the same but with different names, while

R.W. Burry, *Immunocytochemistry*, DOI 10.1007/978-1-4419-1304-3_8,
© Springer Science+Business Media, LLC 2010

some controls use names that do not tell us what is being tested. To simplify the discussion, the controls here are divided into the following three types:

1. 1° antibody controls that show specific binding of the 1° antibody to its antigen
2. 2° antibody controls that show specific 2° antibody to the 1° antibody
3. Label controls that show labeling results from the labels attached to the antibodies

1. 1° Antibody Controls

The 1° antibody control is a specificity control that confirms that the 1° antibody did bind to the correct epitope on the expected antigen. Antibodies bind to epitopes on antigens, and each antigen can have multiple epitopes (Chapter 4, Antibodies). In addition, single epitope can be found on different antigens. This distinction between epitope and antigen is important in understanding the specificity of the 1° antibody.

In immunocytochemistry, binding of the 1° antibody (Fig. 8.1a) is assumed to be to one epitope located on only one antigen (Fig. 8.1b). However, if a 1° antibody can bind to an epitope found on more than one different antigen (Fig. 8.1c, antigens No. 3 and No. 5), then the 1° antibody is not specific.

The four important 1° antibody specificity control methods are listed here in the order of preference among experienced users.

The best control for the 1° antibody is to compare tissue sections from a normal animal and a *knockout animal* in which the same antigen protein is missing (genetic approach). This method is preferred because it shows the antibody label absent after the same incubation conditions used for immunocytochemistry (Fig. 8.2a). Sometimes the knockout might not be complete, and very low levels of protein expression might be detected with immunocytochemistry. To confirm the knockout, the metabolic effects of the knockout protein must be eliminated, indicating that the protein is not synthesized, even at low levels. It is also necessary to use an animal without the knockout as a positive control. Knockout animals are not available for most antigens, so this control, even though it is preferred, is not used frequently.

Immunoblot (western blot) provides a good demonstration of 1° antibody specificity because it results in a unique, labeled band that confirms the molecular weight of the protein. Immunoblots that reveal a series of bands with a single antibody show that the antibody binds to multiple proteins and is not specific. However, the immunoblot does not test for the conditions where the tissue has been fixed and processed for immunocytochemistry. Identification of a single band with the correct molecular weight shows the specificity for an antigen, but unfortunately, the processing of immunoblots uses a denatured protein and not a fixed and detergent-treated protein. In fixed tissue, a protein is likely to retain its quaternary structure and antibodies will be able to bind to folded amino acid sequences that are some distance apart. In immunoblot, the proteins are denatured or linearized; thus, antibodies will mainly recognize contiguous sequences of amino acids and not a folded structure. One approach to make this type of control more like

Fig. 8.1 1° antibody-binding antigens. (**a**) Five different antigens are shown, but the 1° antibody was made to bind to just one of these antigens. (**b**) The 1° antibody can bind only to the correct antigen No. 3. (**c**) A second antigen can contain the same epitope as the antigen used to generate the 1° antibody. In this case, antigens No. 3 and No. 5 label even though only antigen No. 3 was correct

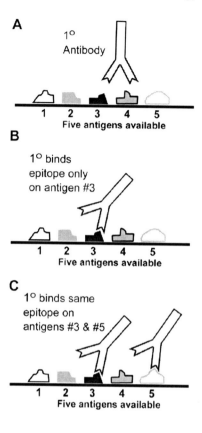

immunocytochemistry is to use immunoprecipitation with the primary antibody of solubilized proteins. This sample is still not fixed, but it will have a configuration more like that of the protein in the tissue. After immunoprecipitation, the sample is run on a gel and silver stained to check for the correct molecular weight. The immunoblot is commonly used because it is easy to perform and shows, albeit by molecular weight separation, that the antigens are different.

Comparison of immunocytochemical localization of the 1° antibody binding with known antibody or label for the correct protein or colocalization uses an immunocytochemical method of tissue sample preparation to determine whether the labeling is the same as the 1° antibody.

Easiest are sections from transgenic animals with GFP tagged to the protein, which shows colocalization (Fig. 8.2b). A second approach is to use a second 1° antibody made to the same antigen, which should show colocalization. In many cases, antibodies to different parts of the amino acid sequence on a single protein antigen use different epitopes on the same antigen and should show 100% colocalization (Fig. 8.2c). This is indirect evidence because it does not show the identical protein bound by the 1° antibody and other antibodies; it only shows the same location is labeled but not necessarily the same protein.

Fig. 8.2 1° antibody
specificity controls. The
specificity of the 1° antibody
for the antigen is
demonstrated by several
methods. (**a**) Knockout
control antigen No. 3 is not
present. (**b**) Colocalization of
a GFP-tagged protein and a
1° antibody to that protein.
(**c**) Colocalization of an
antigen with two different 1°
antibodies

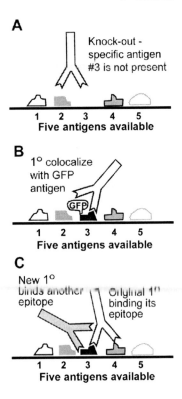

Adsorption controls use isolated antigens to bind the 1° antibody so that when
the adsorbed antibody is incubated with tissue, it can no longer bind the antigen
in the tissue (Fig. 8.3a). In theory, the 1° antibody will bind the isolated antigen
and the bound 1° antibody will not be able to bind the antigen in the tissue. These
controls are difficult to plan and perform, potentially giving false negatives and false
positives (Burry, 2000).

Adsorption controls require titration of the antigen to show that at low antigen
concentrations, the antigen partially blocks the labeling by the 1° antibody, and
at high concentration, the antigen completely blocks labeling. Perform adsorption
controls over a range of concentrations of antibody and antigen to insure that the
correct ratio of antigen to antibody is achieved, and that at high concentrations,
there is an excess of the antigen to block all of the antibodies.

An incorrect negative result can occur when the antibody binding to the same
epitope occurs on multiple proteins (Fig. 8.3b). Examples of such false negative
for epitopes occur when the epitopes are shared on several antigens and bind to all
antigens by adsorption (e.g., Swaab et al., 1977). To insure a negative absorption
control is correct, an immunoblot must show that the epitope is found on a single
protein (e.g., Wolf et al., 2001). Thus, the absorption control needs an additional
control to eliminate the possibility that the 1° antibody is bound to multiple antigens.

Fig. 8.3 1° antibody absorption controls. Incubation of the 1° antibody with the antigen used to generate the antibody is used as a specificity control. (**a**) The 1° antibody with excess antigens will bind all of the Fab sites capable of binding the antigen in the tissue (*arrow*). (**b**) If the correct antigen and an incorrect antigen have the same epitope (*arrows*), then binding to both will be inhibited by the absorption control. (**c**) In some cases, the antigen binds to proteins in the tissue and the adsorbed antibody will bind to a protein (No. 4) independently of the antibody

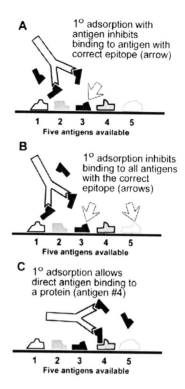

A 1° adsorption with antigen inhibits binding to antigen with correct epitope (arrow)

1 2 3 4 5
Five antigens available

B 1° adsorption inhibits binding to all antigens with the correct epitope (arrows)

1 2 3 4 5
Five antigens available

C 1° adsorption allows direct antigen binding to a protein (antigen #4)

1 2 3 4 5
Five antigens available

Also, false positives can result with absorption controls that occur as part of normal protein–protein binding. An antigen in the absorbed antibody–protein complex still has sites on it that bind to other proteins (Fig. 8.3c). This binding to the cell by normal protein–protein interactions using the attached blocking protein gives an apparent positive result, even though the mechanism does not involve the 1° antibody. When peptides are used to generate antibodies, the limited sequence of the protein will not show this type of false positive.

Finally, adsorption controls can only be performed with purified antigens. In cases of isolated proteins, it is difficult to get a purified antigen clean enough to use as the antigen for an absorption control. For antibodies made from small synthesized peptides, it is possible to get pure antigen and these are the best absorption controls.

Recommendations for 1° Antibody Controls

Best choice is tissue from a knockout animal used with the 1° antibody but these animals are rarely available.

Most commonly used controls are immunoblots with the tissue of interest to determine whether the 1° antibody can bind to a single protein of the correct molecular weight.

A good choice is immunocytochemistry with the 1° antibody and cells labeled for GFP or with a second 1° antibody to the same antigen.

Absorption controls should be used with caution and combined with other controls.

2.2° Antibody Controls

The 2° antibody control, also known as the negative control or the technique control in the experimental context, shows specific binding of the 2° antibody to a 1° antibody (Fig. 8.4a, b).

These controls are easy to perform, but require some experimental planning. In all experiments, omit the 1° or replace the 1° with same species serum at the same dilution, and incubate with 2° antibody, followed by all other parts of the detection method (e.g., ABC, chromogen development). These controls must be performed each time an experiment is performed. If labeling occurs when the 1° antibody is omitted, it must be due to nonspecific binding. Eliminate this nonspecific binding by using of blocking agents (Chapter 5, Block and Permeability).

In some cases, 2° antibody binding to tissue may be due to the lack of specificity of the 2° antibody (Houser et al., 1984; Sterling, 1993). Changing the blocking

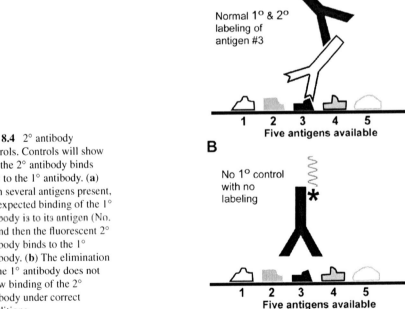

Fig. 8.4 2° antibody controls. Controls will show that the 2° antibody binds only to the 1° antibody. (**a**) With several antigens present, the expected binding of the 1° antibody is to its antigen (No. 3) and then the fluorescent 2° antibody binds to the 1° antibody. (**b**) The elimination of the 1° antibody does not allow binding of the 2° antibody under correct conditions

agents here has minimal effect on the inappropriate labeling, but a more specific 2°
antibody produces better labeling. Purchase 2° antibodies from reliable vendors. Be
sure the antibodies are purified tested in the user's lab. The product literature will
state that the antibody is screened against other species (e.g., human) IgG. For 2°
antibodies used with mouse 1° antibodies, frequently the subclass of 1° antibody
needs to be considered in selecting a 2° antibody. The 2° antibody needs to be made
to bind to the subclass of the IgG used as the 1° antibody.

In experiments with multiple 1° antibody and multiple 2° antibodies, the control
must show that (1) each 2° binds with only the correct 1° antibody and (2) each
2° does not bind to other 2° antibodies. Detailed protocols for multiple 1° antibody
experiments will vary depending on the number of 1° antibodies, the species used
to make the 1° antibodies, and the labeling protocol used. Descriptions of multiple
antibody methods in Chapters 11 and 12 will include a discussion of the specific 2°
antibody controls needed.

3. Labeling Controls

Labeling controls or detection controls help the researcher identify the contribution
of endogenous fluorescence or enzyme to the observed label. Although it seems
unlikely, the label (fluorescence or enzyme) can occur in cells endogenously. For
fluorescent labeling, the control includes incubation of tissue sections or cultured
cells for the needed times with the incubation solutions without antibodies. When
examined with wide field fluorescent microscopy, any fluorescent labeling indicates
endogenous fluorescence background (Fig. 8.5a, *arrow*). For enzyme activity, the
control step should be to incubate the sections or cultures in solutions without
antibodies and perform the chromogen development step. When these are examined with bright field microscopy, the presence of any reaction product indicates
background enzymes (Fig. 8.5d, *arrow*).

When endogenous fluorescence is detected, it can be blocked (Fig. 8.5b, black
"X"). *Autofluorescence* is fluorescence observed in fixed cells without incubation in
any fluorescent compound (Billinton and Knight, 2001). An important characteristic
of autofluorescence is that it has a broad range of wavelengths for both excitation
and emission, which makes it difficult to work around when using other standard
fluorophores. The appearance of autofluorescence in images has one of two general
patterns: (1) irregular or particulate autofluorescence (elastin, lipofuscin, NADH,
flavins, chlorophyll, hemaglobin, etc.) or (2) diffuse or uniform autofluorescence
caused by aldehydes in formalin-fixed tissue and glutaraldehyde-fixed tissue (not a
problem in paraformaldehyde-fixed tissue).

There are many methods for potentially reducing autofluorescence. The key word
is reducing rather than eliminating. Listed here are some methods in decreasing
order of effectiveness. First, increase the signal from the 1° antibody so that the
signal-to-noise (specific label to background autofluorescence) ratio allows the label
to be seen more strongly. Second, most autofluorescence emission spectra are in
the visible green and red ranges, so shifting the specific label to high red labels is

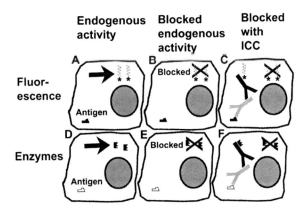

Fig. 8.5 Labeling controls. It is possible that the tissue naturally contains the same label that is attached to the antibody. (**a**) Endogenous fluorescence is present in the cell (*arrow*) along with the antigen. Examining sections with no antibody incubations will show the endogenous labeling. (**b**) Blocking the endogenous labeling with chemicals. (**c**) With fluorescent immunocytochemistry, the antibody binding to the antigen will generate the fluorescence and the endogenous fluorescence will not be seen. (**d**) The presence of endogenous enzymes can be detected with chromogen development in sections with no antibody incubations. (**e**) Endogenous enzymes blocked with chemicals. (**f**) With enzyme-based immunocytochemistry, only the label from the antibodies will be seen when the endogenous enzymes have been blocked

effective. Third, fluorescence can come from free aldehyde groups of fixatives (formalin and glutaraldehyde) that can be reduced with incubations in either a blocking amino acid such as glycine or incubation in sodium borohydride. Fourth, there is a waste product, lipofuscin, called the "aging pigment," which is the breakdown product of red blood cells. Lipofuscin granules in cells can be reduced by treating with Sudan Black or $CuSO_4$ (Schnell et al., 1999). Fifth, it is tedious but possible to scan cells and tissues on a fluorescence microscope at the excitation maxima with either confocal or fluorescent light before the immunocytochemistry procedure (Newman and Gabel, 2002). Finally, spectral imaging can be used on the fluorescent-labeled slide to find autofluorescence and subtract it from the spectra of the image. In this sixth method, a series of images is collected at different wavelengths and software separates the different fluorophores and the autofluorescence. This method is available with spectral confocals or the spectral camera (van der Loos, 2008). In reducing autofluorescence, avoid the use of image analysis software to manipulate images after collection because this changes the specific signal as well as the autofluorescence levels and decreases the image quality. Once blocked, then the fluorescence activity associated with the 2° antibody will be the only thing detected (Fig. 8.5c, black "X").

Endogenous peroxidase activity is seen in many cell types, often due to red blood cells or even white cells that remain in the tissue. Another cause is peroxidase activity in membrane-bound organelles of the lysosome family (Fig. 8.5d), which are found particularly among in phagocytic cells at injury sites. Treating tissue before immunocytochemistry can efficiently eliminate endogenous enzyme activity (Fig. 8.5e), allowing HRP to be used. Several methods are described below that can be used to reduce or eliminate endogenous peroxidase. Try these methods on cells or tissues to determine which works best for the specific application: (1) Pretreating with hydrogen peroxide, with or without methanol, is most effective in inhibiting endogenous peroxidase. (2) Pretreating with phenylhydrazine inhibits peroxidase activity. (3) Using a different enzyme, alkaline phosphatase, instead of HRP as the label. Once blocked the enzyme activity associated with the 2° antibody will be the only thing detected (Fig. 8.5f, black "X").

Protocols for Autofluorescence

Sodium borohydride reduction of autofluorescence (Beisker et al., 1987; Clancy and Cauller, 1998). Note that sodium borohydride is a highly reactive and potentially explosive compound, handle with care maintain a dry environment, and avoid contact with water.

(1) Fix tissue or cultured cells.
(2) Rinse to remove fixative.
(3) Mix 1 mg/ml of sodium borohydride in PBS and use immediately for 10 min. Be careful, vigorous bubbling will cause tissue and coverslips to float.
(4) Rinse in distilled H_2O four times.
(5) Process for immunocytochemistry.

Cupric sulfate reduction of autofluorescence (Schnell et al., 1999)

(1) Follow all immunocytochemistry steps.
(2) Treat sections with solution of 10 mM $CuSO_4$ in ammonium acetate buffer (50 mM; pH 5.0) for 15 min.
(3) Rinse briefly in distilled H_2O five times.
(4) Mount immediately. The cupric sulfate must remain in the tissue during viewing in the microscope.

Sudan Black reduction of autofluorescence (Schnell et al., 1999; Baschong et al., 2001)

(1) Prepare 1.0% Sudan Black in 70% ethanol and stir for 2 h in the dark.
(2) Follow all immunocytochemistry steps, incubate tissue for 10 min.
(3) Rinse briefly in PBS three times and in distilled H_2O four times.
(4) Mount immediately.

Use light to scan cells and tissues at the excitation peak using confocal or using fluorescent light (Newman and Gabel, 2002).

Spectral imaging can identify autofluorescence and subtract it from the spectra of the image either with spectral confocals or the spectral camera (van der Loos, 2008).

Protocols for Endogenous Peroxidase

Hydrogen peroxide

(1) Mix 0.3% H_2O_2 (10 μl 30% H_2O_2 in 1 ml) in PBS.
(2) Incubate with cells or tissue for 30 min before immunocytochemistry.
(3) Rinse four times in PBS.

Hydrogen peroxide and methanol

(1) Mix 0.3% H_2O_2 (10 μl 30% H_2O_2 in 1 ml) 70% methanol in PBS.
(2) Incubate with cells or tissue for 30 min before immunocytochemistry.
(3) Rinse four times in PBS.

Phenylhydrazine

(1) Mix 0.014% phenylhydrazine in PBS.
(2) Incubate with cells or tissue for 30 min before immunocytochemistry.
(3) Rinse four times in PBS.

Chapter 9
Method and Label Decision

Keywords Immunohistochemistry · Antibody labeling · Fluorescence microscopy · Fluorescent immunocytochemistry · Fluorescent immunohisto-chemistry · Indirect immunocytochemistry · Immunostaining

Contents

Introduction 89
Choose Application Label and Method 89
Experimental Design Chart 93

Introduction

In planning an immunocytochemistry experiment, decisions must be made about the exact method needed. The possibilities discussed in this chapter support a complete planning strategy for the experiment. Initially, the scientist will explore different application methods with different label types, then select for a single method and a label. The next step in the planning process is determining the details of finding and preparing reagents. The Immunocytochemistry Experimental Design Chart (included in this chapter) lists the choices that drive the method selection. Filling out the chart helps guide decisions relative to the details of the immunocytochemistry experiment and pin down the final procedure. Generating the step-by-step procedure for the experiment follows in the next chapter (Chapter 10, Single Antibody Procedure). Decisions made based on the chart in this chapter are needed before designing a first immunocytochemistry procedure.

Choose Application Label and Method

The first two design steps in the immunocytochemistry experiments are (1) whether to use fluorescence or enzyme-based labels and (2) which application method is

R.W. Burry, *Immunocytochemistry*, DOI 10.1007/978-1-4419-1304-3_9,
© Springer Science+Business Media, LLC 2010

needed. To make these decisions, consider how protocols to detect primary antibodies relate to the *detection resolution* and *detection sensitivity*. Microscopy is an
important part of the decision. If fluorescence microscopy is chosen then either a
wide field fluorescence microscope or a confocal microscope needs to be available.

Detection resolution measures the degree to which the label localizes to the 1°
antibody site in the cell. The highest detection resolution occurs when the label
is bound directly to the 1° antibody. Direct immunocytochemistry with fluorescent label uses no antibodies or molecules to bridge the label to the 1° antibody.
Graphically, the highest detection resolution occurs with the label attached directly
to the 1° antibody (Fig. 9.1a, *arrow*).

Low resolution occurs when enzymes (i.e., HRP), localized in the microscope
generate a reaction product that spreads from the site of the enzyme and decreases
the detection resolution. An example of lowest detection resolution would be ABC
immunocytochemistry, because an enzyme is the label and it is bridged to the
primary antibody by avidin–biotin (Fig. 9.1a, number 2, Low resolution). Any additional molecules that bind to the 1° antibody and bridge the label away from the 1°
antibody will decrease the detection resolution. In planning experiments, resolution
is important when antigens are localized to a discrete location.

Detection sensitivity measures the intensity of label in the microscope following
the immunocytochemical procedure. Detection sensitivity is a direct measure of the
amount of label at the 1° antibody location or the amount of label generated in the
case of an enzyme. An example of high detection sensitivity is ABC immunocytochemistry, where a large amount of HRP reaction product is generated by a single

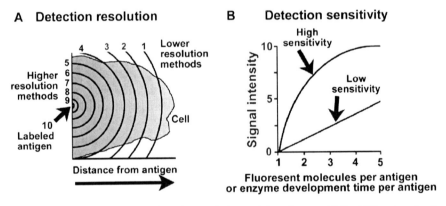

Fig. 9.1 Immunocytochemistry detection resolution and detection sensitivity. Different methods
of applying antibodies and different label types for immunocytochemistry have a range of properties. To select a detection method and a label evaluate these properties according to individual
experimental needs. (**a**) Detection resolution is a measure of the distance between the 1° antibody-
binding antigen label and farthest extent of the label. Schematically, it is represented by concentric
circles from the antigen site, with 10 being the highest resolution and 1 the lowest. (**b**) Detection
sensitivity is a measure of the intensity of the label for one antibody bound to an antigen. High
sensitivity has saturation of detection with only a few label molecules for one antigen, while low
sensitivity has low levels of label, even with many label molecules for one antigen

enzyme (Fig. 9.1b, *arrow* high sensitivity). An example of low detection sensitivity is fluorescence, where the emission from a few fluorophores is difficult to detect (Fig. 9.1b; *arrow* low sensitivity). In planning experiments, the detection sensitivity is most important with antigens that are infrequent in cells or are diffused within cells. Note, the detection sensitivity is only a measure of label detection and is not related to the specificity of the 1° antibody to bind the antigen.

To help evaluate possible labels and methods, use Table 9.1. Note that the number ratings in Table 9.1 are based on the author's experience and should only be used to compare methods within this table. Detection resolution is the ability to localize the label to the exact site of the primary antibody. The direct immunocytochemistry method does this and is rated as a detection resolution of 10 (Table 9.1, Fig. 9.1a, *arrow*). With detection methods that have lower detection resolution the number decreases (Table 9.1, Fig. 9.1a, numbers).

Importantly, there is an inverse relationship between the detection sensitivity and the detection resolution. In general, the most sensitive method has the greatest spread of label. For example, localizing a protein to the *cis* part of the Golgi apparatus with fluorescent indirect immunocytochemistry (detection resolution = 9) will give a discrete label, but if the HRP direct avidin–biotin immunocytochemistry (detection resolution = 4) were used, the label would fill the region of the Golgi apparatus and possibly the entire cell. On the other hand, if the protein could be only weakly detected with the indirect immunocytochemistry method (detection sensitivity = 4), then the avidin–biotin immunocytochemistry direct with enzyme could detect it strongly (detection sensitivity = 8).

Table 9.1 Detection sensitivity and detection resolution of application methods

Application method	Label types	Detection resolution	Detection sensitivity	Number of incubation steps*
Direct immunocytochemistry	Fluorescent	10	1	1
Indirect immunocytochemistry	Fluorescent	9	3	2
TSA immunocytochemistry	Fluorescent	8	5	3
Avidin–Biotin immunocytochemistry – Direct	Fluorescent	7	6	3
Indirect immunocytochemistry	Enzyme	4	7	3
Avidin–Biotin immunocytochemistry – Direct	Enzyme	3	8	4
ABC immunocytochemistry	Enzyme	2	9	5
Combine ABC and TSA immunocytochemistry	Enzyme	1	10	7

Detection sensitivity and detection resolution are measured on a 1–10 scale, with 10 being best and 1 worst
*Number of Incubation Steps does not include rinses.

In another example, if the goal is to locate neuronal cell bodies containing a specific neurotransmitter, a high detection sensitivity and low detection resolution method would allow the label to diffuse throughout the entire cell. Then the location of specific neurotransmitter molecules within the cell is not important; rather location of a cell that contains neurotransmitter is important. The bottom line is to assess the importance of detection resolution and detection sensitivity in selecting a detection method.

To guide the application method for an experiment, consider the answers to two questions:

What detection sensitivity is required? Start by deciding the frequency of antigens of interest in the cells being examined. That is, are the antigens common or rare? Does the antibody have good sensitivity for the antigen? In most cases, if the antigen is relatively common and the antibody has good sensitivity, a detection method with relatively low sensitivity will work. If the antigen is rare or the 1° antibody binds weakly to the antigen, a detection method of high sensitivity will be needed. So why not use high-sensitivity methods all the time? Unfortunately, high-sensitivity methods come at a cost; they have many more steps (places for errors) and lower resolution for detecting the label in the microscope. The best advice is to use only the level of sensitivity needed to give results. In most cases, the level of detection resolution drives the selection.

What detection resolution is needed? Consider the experiment and the need for localization of label to a specific organelle, nucleus, or cell. For localization to a vesicle, methods with detection resolutions of seven or higher will be needed, while localization to a cell could use detection resolutions as low as 1. A method with low or poor detection resolution results in a label that covers a larger area than the location of the 1° antibody. With high detection resolution, the label approaches the size of the 1° antibody. There is an inverse relationship between the detection resolution and the detection sensitivity. A method with high detection resolution will have low detection sensitivity.

Answers to the questions should point to a label type and a method of application for the antibodies. For high-resolution samples, generally the needed label is fluorescence. For lower-resolution samples, both fluorescence and enzymes can be used. However, before deciding on a label type, determine whether the correct microscope is available. A wide field fluorescence (epifluorescence) microscope or a confocal microscope is required for fluorescent type of label. Also, check the fluorescence microscope to determine whether it has the correct filter sets (Chapter 13, Microscopy and Images).

Finally, consider the number of steps that are feasible for the planned experiment because steps reflect the complexity of the experiment. The number of steps will be largely determined by the answers to the questions. Simple experiments involve two antibody incubations, each followed by four buffer rinses for a total of ten steps over 1 day. Complex experiments can involve six to eight antibody incubations, each followed by four buffer rinses for a total of 30 steps over 3 days. Consider the number of steps as a way to assess whether the complexity of the selected method is appropriate. More steps means an increased chance for mistakes and added expense to the experiments.

Experimental Design Chart

Experimental Design Chart guides the selection of antibodies, incubation conditions, and solutions in planning an immunocytochemical experiment. After selecting the method of antibody application and selecting the label, the next step is to select and collect the needed reagents. The decisions modeled here are based on material that was explained previously in this book. To help in the decision process, an example immunocytochemistry Experimental Design Chart (Table 9.2) is included to help in selecting reagents.

The chart is divided on the left into six categories that should be followed from the top down to complete the chart for each experiment. The Appendix contains chart templates for various methods of antibody detection presented in this book. The second column from the left lists parameters within each category where specific information is needed. The boxes surrounded by dark lines on the right side of the chart (filled in this example) indicate two types of information needed. The column with the heading, Conditions, collects information about the procedure independently of the antibodies to be used. The last column, Antigen No. 1, asks for information specific for each antibody to be used. Most of the experiments scientists perform use at least two antibodies and the chart allows for one antibody; additional columns can be added for more antibodies.

The first category (Table 9.2), *(1) Sample,* collects the conditions of sample preparation. In the example here, kidney from a rat will be used following *trans*-cardiac perfusion. The tissue blocks will be infiltrated with 20% sucrose overnight, frozen in isopentane, and sectioned on a cryostat. For processing, the sections will be placed on a microscope slide. To calculate how much incubation and rinse solutions are needed, the size of the incubation chamber is listed. In the example, 250 µl will just cover the tissue and can be used for each step.

In the second category, 1° *antibody*, antigens are listed in separate columns so that all of the reagents associated with the 1° antibody can be seen. In this example, the 1° antibody to antigen Ag A is made in mouse. The dilutions of the 1° antibody for anti-Ag-A is not known, as this antibody has not be used previously, so the dilution will be determined by the Dilution Matrix when the procedure is developed in Chapter 10 (Single Antibody Procedure).

The third category, 2° *antibody*, summarizes 2° antibody and label information. This information allows the scientist to check the species of the 2° antibodies so that no cross-reactive antibodies will be used. In addition, the information about the fluorophores lists the excitation and emission wavelengths to show that different nonoverlapping fluorophores will be used.

The fourth category, *Incubation Solutions,* includes information on the species of the normal blocking serum and the detergent.

Table 9.2   Experimental design chart

Experimental design chart
Indirect immunocytochemistry

Category	Parameter	Conditions	Antigen No. 1
(1) Sample			
	Source and tissue	Rat kidney	
	Fixative	4% paraformaldehyde	
	Fix application method	Perfused	
	Embedding	20% sucrose over night and isopentane freezing	
	Sectioning	Cryostat	
	Incubation chambers/size	Microscope slide 100 ml per area	
(2) 1° Antibody			
	Antigen		Ag A
	Source of 1° antibody		Abs are Us No. 123
	Species of 1° antibody		Mouse anti-Ag A
	Dilution 1° antibody		Not known
	Mixing 1° antibody		Not known
(3) 2° Antibody			
	Species		Goat anti-mouse
	Source of 2° antibody		Abs are Us No.987
	Fluorophore		Fluo 488
	Excitation wavelength		488 nm
	Emission wavelength		505 nm
	Dilution 2° antibody		Not known
	Mixing 2° antibody		Not known
(4) Incubation solutions			
	Buffer	PBS pH 7.2	
	Block serum species	5% normal goat and 1 mg/ml BSA	
	Detergent	0.3% Triton	
(5) Controls			
	1° antibody controls		Done immunoblot product lit.
	2° antibody controls		Need to be done
	Label controls		Done no fluorescence
(6) Microscope			
	Location	Confocal in core microscope facility	
	Fluorescent filters		488 nm band pass
	Lasers		Argon

The fifth category, *Controls*, details decisions about the three different types of controls. In this example, the 1° antibody controls show that the specificity was determined by the company supplying the 1° antibodies. In this case, the company showed immunoblots with the antibody in the product literature. In some cases, it might be necessary to repeat the specificity control. The 2° antibody controls must be included each time the experiment is performed and consist of replacing the 1° antibody with the same dilution of normal serum from the same species of animals. The 2° antibody controls might be more extensive when multiple primary antibodies are used. In the example here, label controls were done on this tissue previously by examining a mounted section with no labeling using the same confocal microscope.

The sixth category, *Microscopy*, lists the microscope needed to examine the sections at the end of the experiment. Especially when using new labels and/or a new microscope, check that the correct fluorescent filters and illumination are available. If the results are to be collected on a confocal microscope, the correct filter set and laser must be available.

Completing the Immunocytochemistry Experimental Design Chart guides decisions about the reagents need for the experiment. Chapter 10 (Single Antibody Procedure) presents the design strategy based on the individual steps needed for each antibody.

Chapter 10
Single Antibody Procedure

Keywords Immunohistochemistry · Antibody labeling · Fluorescence microscopy · Fluorescent immunocytochemistry · Fluorescent immunohisto-chemistry · Indirect immunocytochemistry · Immunostaining

Contents

Introduction 97
Experimental Design Chart 98
Incubation Conditions 98
Antibody Dilutions 100
Antibody Dilution Matrix 102
2° Antibody Controls 102
Rinses 104
Mounting Media 105
Final Procedure 106
   Steps in a Single 1° Antibody Indirect Immunocytochemistry Experiment 106
   Steps in a Single 1° Antibody Immunocytochemistry Experiment for Ag A 107

Introduction

Designing an experiment with a single 1° antibody involves many of the topics presented in the previous chapters, including reagents, types of label for antibod-ies, and the methods for applying antibodies. This chapter presents a step-by-step approach to designing an immunocytochemistry experiment and depends on infor-mation in the Immunocytochemistry Experimental Design Chart completed in Chapter 9.

The example procedure presented here with rat kidney detects the antigen Ag A (Fig. 10.1a), with a mouse anti-Ag A antibody (Fig. 10.1b), and a goat anti mouse 2° antibody labeled with a 488 fluorophore (Fig. 10.1c).

R.W. Burry, *Immunocytochemistry*, DOI 10.1007/978-1-4419-1304-3_10,
© Springer Science+Business Media, LLC 2010

Fig. 10.1 Single 1°
antibody indirect
immunocytochemistry.
(**a**) Antigen Ag A before
incubations with antibodies.
(**b**) The 1° antibody mouse
anti-Ag A binds to Ag A.
(**c**) The 2° antibody goat
anti-mouse 488 fluorophore
binds to the mouse anti-Ag A

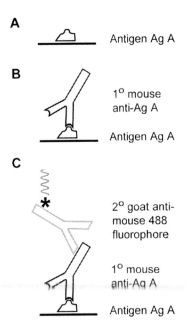

A

Antigen Ag A

B

1° mouse
anti-Ag A

Antigen Ag A

C

2° goat anti-
mouse 488
fluorophore

1° mouse
anti-Ag A

Antigen Ag A

Remember, all experiments, including multiple 1° antibody experiments, begin with single 1° antibody immunocytochemistry. Because performing multiple 1° antibody experiments can be complex due to antibodies that potentially interact, it is absolutely necessary to demonstrate each antibody works singly; thus, begin with single 1° antibody experiments.

Experimental Design Chart

In planning experiments, it is important to look at the entire experiment and know that all of the reagents and conditions have been considered. Table 10.1 is the Experimental Design Chart for a single antibody experiment. The most common use for single antibody experiments is to test a new 1° antibody and to perform the Dilution Matrix described later. Before planning the details needed to perform each of the steps in the experiment, this chart must be completed.

Incubation Conditions

The key with antibody incubations is the penetration of antibodies into the tissue or cells. This is like movement of liquids through a sponge. Water can easily move into and through a sponge because the pore size of the sponge is much larger than the size of the water molecules. If fine sand in water is used, it is difficult to get the sand into

Table 10.1 Experimental design chart

Experimental design chart
Indirect immunocytochemistry

Category	Parameter	Conditions	Antigen No.1
(1) Sample			
	Source and tissue	Cultured fibroblasts	
	Fixative	4% paraformaldehyde	
	Fix application method	Drop in	
	Embedding	20% sucrose over night and isopentane freezing	
	Sectioning	Not applicable	
	Incubation chambers/size	24-well plate	
(2) 1° Antibody			
	Antigen		Ag A
	Source of 1° antibody		Abs are Us No.123
	Species of 1° antibody		Mouse anti-Ag A
	Dilution 1° antibody		Not known
	Mixing 1° antibody		Not known
(3) 2° Antibody			
	Species		Goat anti-mouse
	Source of 2° antibody		Abs are Us No.987
	Fluorophore		Fluo 488
	Excitation wavelength		488 nm
	Emission wavelength		505 nm
	Dilution 2° antibody		Not known
	Mixing 2° antibody		Not known
(4) Incubation solutions			
	Buffer	PBS pH 7.2	
	Block serum species	5% normal goat and 1 mg/ml BSA	
	Detergent	0.3% Triton	
(5) Controls			
	1° antibody controls		Done immunoblot product lit.
	2° antibody controls		Need to be done
	Label controls		Done no fluorescence
(6) Microscope			
	Location	Wide field fluorescence microscope	
	Fluorescent filters		488 nm band pass
	Lasers		Not applicable

the sponge because its pore size is close to the size of the sand. This is like getting antibodies into tissue. The proteins, membranes, and the fixative set up a network of material, similar to the spaces within the sponge. Back to the sponge analogy, the most important property to increase penetration of sand into the sponge is by motion or agitation of the sponge in the sand solution. The motion of the sponge increases the chances that the individual sand grains will move through the tortuous paths into the sponge. A second important property is time of incubation. Given enough time, it is possible to get a considerable amount of sand into a sponge.

The first step before incubations with antibodies is the use of detergents to solubilize the membranes of the cells and tissues. As discussed in Chapter 5, selection of a detergent and its incubation conditions are important. The best penetration of antibodies is found after removal of as much of the cellular membranes as possible.

Incubation conditions for tissues and antibodies use the same ideas. *Agitation* of the samples during incubation will both move reagents through the tissue and help keep the highest concentration of reagents at the exposed surface of the tissue or cells. Agitation is highly recommended and will shorten the incubation times. In fact, not using agitation will sometimes result in irregular antibody penetration, resulting in splotchy labeling or producing uneven penetration with label missing from some areas and too heavy in other areas.

Incubation time allows reagents to diffuse into and through cells and tissue. Ideally, incubation times that cover multiple days would be best. However, the incubation time is usually determined by thickness of the cells or tissue. Most commonly, the incubation time is set to a short time for thin cell cultures (4 h), or longer times for thick tissue sections (days).

Incubation temperature options are either room temperature (23°C) or ice/refrigeration (4°C). While the increased penetration due to higher temperature alone is small, the biggest issue is the effect of temperature on the lipid membranes. At 4°C, membranes are rigid and restrict movement of antibodies. At room temperature, the membranes are more easily penetrated. Incubations at room temperature are preferred because they allow for the best diffusion of reagents into the tissue in the shortest amount of time. The problem with room temperature incubation is bacteria can grow in incubation solutions, and cold temperature incubations are used to inhibit bacterial growth. Incubations at ice/refrigeration temperatures inhibit that bacterial growth, but increase the incubation times with reduced penetration. An alternative for longer incubations at 4°C is the use of 0.02% sodium azide in the buffer that inhibits the growth of bacteria. For fluorescence immunocytochemistry, sodium azide should be a routine part of all buffers. However, for HRP-based immunocytochemistry, sodium azide will inhibit the development of the HRP chromogen and should not be used in the final rinses.

Antibody Dilutions

Determining the optimal antibody dilution is crucial in achieving high specificity, low background, and reasonable cost of antibodies. Remember, "more is not always better" with low antibody dilutions (e.g., 1:50) specific labeling might not be as good as using a higher dilution (e.g., 1:1000). In fact, if the 1° antibody dilution is too low, rather than seeing heavy labeling, almost no labeling might be seen. For each 1° antibody and each 2° antibody, follow *antibody Dilution Matrix* to determine the optimal dilution for both the 1° antibody and the 2° antibody.

Antibody dilutions are ratios. A 1:100 dilution is 1 unit of antibody in 100 units of buffer or 1 μl of antibody in 100 μl buffer. Note, a 1:100 dilution is made by adding 1 μl of antibody to 99 μl buffer for a total of 100 μl. The best way to calculate a dilution is multiply the dilution (1:100) as a fraction (1/100) by the final volume needed (1000 μl) to determine the volume of the antibody to be added (10 μl). A dilution is not expressed in units (e.g., 1:100).

$$\text{Dilution ratio} \times \text{final volume} = \text{Volume of antibody to be added}$$

For a 1:100 dilution in 1 ml (1000 μl):

$$1/100 \times 1000\,\mu l = 10\,\mu l$$

Thus, for a 1:100 dilution in 1 ml (1000 μl) final volume, add 10 μl of stock antibody to 990 μl buffer.

For a 1:2000 dilution in 1 ml:

$$1/2000 \times 1000\,\mu l = 0.5\,\mu l$$

Thus, for a 1:2000 dilution in 1 ml (1000 μl) of final volume, add 0.5 μl of stock antibody to 999.5 μl buffer.

For a 1: 50,000 dilution in 1 ml:

$$1/50,000 \times 1000\,\mu l = 0.00002\ (2 \times 10^{-5}) \times 1000\,\mu l = 0.02\mu l$$

Thus, for a 1:50,000 dilution 1 ml (1000 μl) of final volume, 0.02 μl of stock antibody needed. However, this volume is below the ability to measure accurately. Sequential dilutions are required; the first 1:500 and the second 1:100 for a total dilution of 1:50,000 (multiply the two dilutions 1:500 × 1:100 = 1/50,000).

First $1/500 \times 1000\,\mu l = 0.002 \times 1000\,\mu l = 2\,\mu l$ of stock is added to 998 μl buffer.

Second $1/100 \times 1000\,\mu l = 0.01 \times 1000\,\mu l = 10\,\mu l$ of the 1:500 diluted antibody (first step) is added to 990 μl buffer.

Antibody Dilution Matrix

The Antibody Dilution Matrix (after Hoffman et al., 2008) is needed to determine the correct dilution of antibodies. For a different detection method, the matrix is the same, but the dilutions of the 1° antibody and other reagents may be a different antibody Dilution Matrix has different concentration ranges for different detection systems. To test the antibody Dilution Matrix with indirect fluorescent immunocytochemistry, run a preliminary experiment that uses eight samples. The results of this preliminary experiment will show the dilutions for both the 1° antibody and the 2° antibody. This one experiment will save so much time that its importance cannot overemphasized.

Table 10.2 Antibody dilution matrix

Antibody dilution matrix

		1° antibody dilution			
		1:100	1:1,000	1:10,000	No antibody
2ᵛ antibody dilution	1:100				
	1:1,000				

The results of the antibody Dilution Matrix must be documented with micrographs taken with the identical camera settings for the entire set of conditions, so that accurate comparisons between cells of the matrix can be made. Some micrographs will be overexposed because the dilution is low. The digital micrographs can be pasted into the matrix, as can numerical evaluation of labels (Fig. 10.2). The important result is to see the loss of label from left to right and from top to bottom as this shows the expected trends. It is also important to use the same microscope and recording device (camera or detectors) that will be used in the final experiment. Select a 1° and a 2° antibody concentration that gives the lowest background and reasonable labeling and the greatest difference between the signal (labeled cell) and the background (nonlabeled area). The dilution with the most intense labeling frequently has high background and a narrow difference between signal and background. This difference between signal and background (noise) is more important than the absolute intensity of the signal. The labeling for the 2° antibody at a 1:100 dilution with different 1° antibody dilutions shows not only high specific labeling but also high background for all 1° antibody dilutions and the no 1° antibody (Fig. 10.2a, b, c, d). With the 2° antibody at 1:1000, the different 1° antibody dilutions gave low background labeling and good specific labeling at 1:1000. The ideal dilutions in this matrix are 1° antibody at 1:1000 and 2° antibody at 1:1000.

2° Antibody Controls

It is very important to show that all of the observed labeling is due to the binding of the labeled 2° antibody to the 1° antibody (Table 10.3). This control is the processing

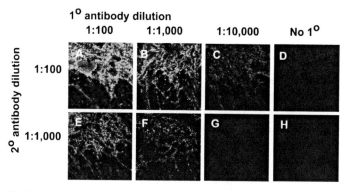

Fig. 10.2 Dilution matrix. An example of micrographs for the Dilution Matrix with a mouse anti-p38 1° antibody and a goat anti-mouse Alexa 546 2° antibody. (**a**) The *top row* of micrographs shows the 2° antibody at 1:100 and begins with a 1° of 1:100 showing high specific and some background labeling. (**b**) With 1° antibody at 1:1000 and the 2° antibody at 1:100, the specific labeling is still saturated and the background labeling is still present. (**c**) With the 1° antibody at 1:10,000 and the 2° antibody at 1:100, the specific labeling is low with background present. (**d**) With no 1° antibody and the 2° antibody at 1:100, background labeling is seen. (**e**) The *bottom row* shows images with the 2° antibody at 1:1000 and begins with a 1° of 1:100 with specific labeling saturated. (**f**) At 1:1000, the 1° antibody and the 2° antibody at 1:1000 specific labeling is good and no background is seen. (**g**) With the 1° antibody at 1:10,000 and the 2° antibody at 1:1000, little specific labeling is seen. (**h**) With no 1° antibody and 2° antibody at 1:1000, no background labeling is seen

Table 10.3 Controls for indirect single antibody

Controls for indirect single antibody

Control type	Conditions	Single 1° antibody
Experimental	1° antibody	Ag A
	2° antibody	Goat anti-mouse 488 fluorophore
	Labeling observed	Normal (green)

No 1° antibody	1° antibody	None
	2° antibody	Goat anti-mouse 488 fluorophore
	Labeling observed	No labeling

of a sample without the 1° antibody incubation, but with the 2° antibody incubation. Perform the 2° antibody control (Chapter 8, Controls) each time the experiments are performed. This requires adding a sample that will not get 1° antibody and will be incubated only with 2° antibody. In the antibody Dilution Matrix, the right column on the right includes the 2° antibody controls. In this control, there should be no labeling and labeling found indicates a problem with the 2° antibody binding to something other than the 1° antibody.

Rinses

The key in *changing solutions* is to remove most, but not all of the solution from the sample. Removing all of the solution risks drying the tissue. When an antibody solution dries on tissues it becomes permanently attached and shows labeling. Plan to change solutions one well at a time to prevent tissue drying. Also, in each well leave enough liquid covering the tissue to prevent drying. When removing a solution, leave 10–20% of the liquid in place at each rinse. Rather than using five rinses of 5–10 min, it is necessary to do at least seven rinses to remove all of the reagents; some investigators use ten rinses!

Rinses after incubations with the 1° antibody before the 2° antibody are important to allow removal of all of the 1° antibody. Incomplete removal of the 1° antibody does not increase background as might be expected rather it lowers the amount of specific labeling (Fig. 10.3). After one rinse (Fig. 10.3a), the amount of specific labeling is very light but detectable (Fig. 10.3c); the third rinse still shows little labeling (Fig. 10.3b), but it is not until the fifth rinse that a reasonable amount of labeling is found (Fig. 10.3c). There is still some increase in labeling at seven rinses (Fig. 10.3d). The explanation for this low labeling with few rinses is the reaction between the 1° antibody and 2° antibody in the solution above the section. The key here is that a reaction in the solution above the tissue section reduces the functional concentration of the 2° antibody to diffuse into the tissue. The 1° antibody in the solution and that coming out of the tissue functions quench the 2° antibody decreasing its concentration. These bound antibody–antibody complexes are washed away with the rinses and are not seen in the tissue. Thus, no increase in background labeling is seen. This concept of antibodies reacting in solution and not being seen in the tissue is important in solving problems that can be encountered in trouble shooting of experiments (Chapter 14).

Rinses after 1° antibody

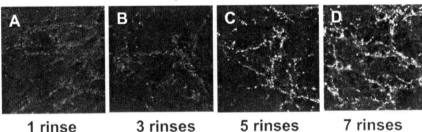

1 rinse   3 rinses   5 rinses   7 rinses

Fig. 10.3 Rinses after the 1° antibody before the 2° antibody. Images show results after normal processing and incubation with the 1° antibody and varying numbers of rinses before a normal incubation with the 2°. (**a**) With one rinse, the specific labeling was low and no background was seen. (**b**) With three rinses, the specific labeling was still low and no background was seen. (**c**) By five rinses, the level of specific labeling increased. (**d**) With seven rinses, the level of specific labeling was high and there was no background. Mouse spinal cord was processed as in Fig. 10.2. The wells for the incubations were 80–100 µl in volume

Rinses after incubations with the 2° antibody prepare the tissue for the mounting medium and insure that no free labeled 2° antibody is present. Incomplete removal of the labeled 2° antibody will lead to increased background (Fig. 10.4). Following 2° antibody incubation, adding mounting medium directly with no rinses will leave small droplets of labeled 2° antibody in the mounting medium (Fig. 10.4a, *arrows*). The level of label in the tissue that is not specifically labeled is high and makes the specific label look weak. After one rinse, the amount of background label is still high (Fig. 10.4b). With three rinses, the background begins to decrease and the specific labeling is seen clearly (Fig. 10.4c). After five rinses, the background labeling is completely removed and only specific labeling is seen (Fig. 10.4d).

Rinses after 2° antibody

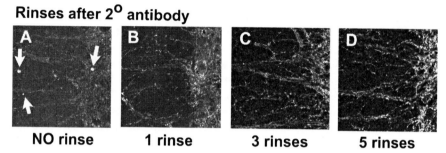

NO rinse   1 rinse   3 rinses   5 rinses

Fig. 10.4 Rinses after the 2° antibody before mounting a coverslip. Images show results after normal processing and incubations with 1° and 2° antibodies, and varying number of rinses before adding the mounting medium. (**a**) Directly after the incubation with the 2° antibody, mounting medium was put on the section. Specific labeling was present and the background was very high. *Arrows* show small droplets of fluorescent 2° antibody. (**b**) Following one rinse, the background was reduced and the specific labeling was good. (**c**) After three rinses, the background was very low. After five rinses, the background was gone and the specific labeling was clearly seen. Mouse spinal cord was processed as in Fig. 10.2. The wells for the incubations were 80–100 μl in volume

In planning experiments, rinses between antibodies must be able to remove all of the previous antibody so that the next antibody will not react with it in solution before it has a chance to enter the tissue. A result of not enough rinsing between antibodies will be reduced labeling rather than increased background. Rinses of the final labeled antibody remove the unbound label and allow the specific label to be seen at highest intensity. If some of the labeled antibody is left in the tissue, the result will be background labeling.

Mounting Media

Following the final rinse, the cells or tissues must be held in place between the microscope slide and a thin (about 100 μm) coverslip that are held in place by mounting media. The refractive index of the mount media is important. The refractive index is a measure of the bending of light when it passes across a boundary between the media such as air and water. The refractive index of water is 1.333, glass has a refractive index of 1.520, and air has a refractive index of 1.000. Because the tissue sections with fluorescent dyes must be mounted in an *aqueous mounting medium*, they need to have the same refractive index of glass.

The original aqueous mounting medium was 80% glycerol in water, which did not harden, and required a rim of finger nail polish around the edge to seal in the liquid. Today, there are many aqueous mounting media that will harden after 24 h at room temperature and do not require a finger nail polish seal. To use these hardening aqueous mounting media, replace all of the buffer solution with the embedding media with a single rinse of media. Aqueous embedding media are Immumount from Thermo Fisher Scientific, Mowiol 40–88 (powder) Sigma, Crystal/Mount or Gel/Mount from BioMedia, and Prolong Gold from Molecular Probes/Invitrogen. Coverslips must be both firmly attached and completely dry before examining under a microscope. Examining a microscope slide with wet mounting media or wet finger nail polish risks the chance of touching the objective lens of the microscope with the wet solution that will harden permanently on the lens. Removing hardened mounting media from an objective lens is difficult!

Mounting media can contain antibleaching agents to prolong the life of fluorescence. Bleaching or loss of fluorescence due to exposure to light is a problem (Chapter 6, Labels). Antibleaching agents such as Vectashield® will reduce the photobleaching common with early fluorescent labels (e.g., FITC). After a few weeks, many of these agents in the mounting medium are oxidized and develop their own fluorescence, which obscures the fluorescent label. With the new fluorescent labels for antibodies (e.g., Alexa Fluor, Molecular Probes/Invitrogen), photobleaching is not a major problem and use of antibleach agents is not necessary.

Final Procedure

The final procedure can now be designed based on the specific requirements of the experiment. The example used here is an experiment design from Table 10.1 above, a single-label immunocytochemistry experiment for Ag A in rat kidney tissue. In this chapter, the goal is to design a step-by-step procedure for each 1° antibody separately.

Steps in a Single 1° Antibody Indirect Immunocytochemistry Experiment

List of Steps

(A) Prepare animal
(B) Fix tissue
(C) Embed, section, and mount tissue
(D) Block and permeabilize

(E) Rinse after block and permeabilize
(F) Incubate 1° antibody
(G) Rinse after 1° antibody
(H) Incubate 2° antibody
 (I) Rinse after 2° antibody
(J) Mount coverslip
(K) Examine in microscope
(L) Evaluate results

Steps in a Single 1° Antibody Immunocytochemistry Experiment for Ag A

Materials List

Animal
Perfusion apparatus
4% Paraformaldehyde in phosphate buffer fix
20% Sucrose in PBS
Styrofoam and beaker for freezing tissue
Aluminum foil
Isopentane
OCT
Freezer –80°C for storing tissue
Cryostat with several chucks
Coated microscope slides
Phosphate buffered saline (PBS)
PBS plus 10 mg/ml BSA, 5.0 % normal goat serum, and 0.02% sodium azide
Block permeabilize solution PBS plus with 0.3% Triton
Mouse anti-Ag A 1° antibody
Goat anti-mouse 488 fluorophore
Rocker or orbital shaker
Mounting medium
Confocal microscope in core facility

List of Steps – Specific Details

(A) *Prepare animal* – For this example experiment, a rat is deeply anesthetized as prescribed by the animal use protocol for the laboratory.
(B) *Fix tissue* – Perfuse the rat with 4% paraformaldehyde in phosphate buffer. After the animal is stiff to the touch, remove the kidney and place in fixative for dissection. Use a new sharp scalpel blade to cut the kidney, which is then submerged in the fixative solution for an additional 2 h.

(C) *Embed, section, and mount tissue* – Infiltrate the tissue blocks in 20% sucrose in buffer prior to freezing. In a cold room at 4°C, place the blocks on an orbital shaker or a rocker overnight. The next morning when the tissue has sunk to the bottom of the tube, it is infiltrated. Freeze sucrose-infiltrated blocks of tissue on strips of aluminum foil in isopentane. A Styrofoam shipping box can serve as a Dewar, which is filled halfway full with liquid nitrogen. Fill a small plastic beaker half full with isopentane and place it in the liquid nitrogen. The liquid nitrogen will cool the isopentane to its freezing point, but can be melted with a warm piece of metal (like a blade of a screwdriver). Quickly, remove the individual tissue blocks from the sucrose solution, blot them, and place each block on the end of a strip of labeled aluminum foil. Plunge each strip into the cold liquid isopentane. Store the frozen tissue on aluminum foil in individual tubes at –70°C until needed. For sectioning on a cryostat, first freeze chucks in a cryostat. Next, place a small amount of cold OCT liquid on a chuck, quickly remove the sample from the aluminum foil, and orient it properly in a freezing OCT-covered chuck. Surround the block with cold OCT and when frozen the tissue is ready for sectioning. Operation of a cryostat for obtaining sections was covered in Chapter 4. This includes making wells on the microscope slide with a hydrophobic liquid pen.

(D) *Block and permeabilize* – Blocking and permeabilization solution contains 10 mg/ml BSA, 5.0% normal goat serum to block nonspecific antibody binding, and 0.02% sodium azide is used to prevent bacterial growth during incubations all in PBS. Perform permeabilization with 0.3% Triton. Incubate for 1 h. Permeabilization is done for a set time during this step and detergent is not used for the incubation steps. As discussed in Chapter 5, Block and Permeabilization, continued exposure of the tissue to detergent during antibody incubations could progressively extract more from the tissue or effect the antibody binding. To make the procedure reproducible it is recommended that the detergent only be used for a set period of time in the blocking and permeabilization step.

(E) *Rinse after block and permeabilize* – Use rinse buffer solution containing 10 mg/ml BSA and 5.0 % normal goat serum in PBS (no detergent) for ALL rinses. To prepare this solution, multiply the number of samples by the volume of each rinse, times the total number of rinses for the entire experiment. For example, if an experiment includes ten samples, we allow 500 μl for each rinse and a total of eight rinses (two rinses after block and permeabilize; six rinses after 1° antibody; six rinses after 2° antibody). A total of 10 samples × 14 total rinses × 0.5 ml per rinse = 70 ml of rinse buffer plus 5% excess is 3.5 ml for a total of 73.5 ml. The excess amount is a safety factor, in case of spill or other unexpected event. The rinses after the block and permeabilize step remove the Triton, limiting its activity to a specific period of time. These rinses are for 5 min with the same agitation used during the block and permeabilize step.

(F) *Incubate 1° antibody* – In this example experiment, the anti-Ag A 1° antibody has not been used in the lab and therefore its dilutions had not been determined. Using the Antibody Dilution Matrix described in this chapter, the conditions for this 1° antibody were found to be mouse anti-Ag A, the dilution is 1:500 and the goat anti-mouse 488 fluorophore, the dilution is 1:1000. This experiment here will have ten samples including experimental samples and control sample. These are nine samples with both 1° and 2° antibody and one 2° control with no 1° antibody and only 2° antibody. To mix the 1° antibody, mouse anti-Ag A at 1:500, use 250 μl per sample. For nine samples that will be 2.25 ml buffer plus 10% = 2.5 ml of buffer and 5 μl of 1° antibody. The incubation with the working 1° antibody solution is done for 6 h because with the 10 um tissue sections, the rocker used at room temperature will give complete penetration through the section.

Some labs attempt to *save money* on antibodies by mixing small volumes that do not allow enough volume to cover the samples or by reusing antibodies multiple times. These cost-saving attempts lead to repeating experiments. Do the math, the amount of money saved on the antibody is much less than the hourly rate paid to prepare the animals and do the experiment. Do not cut corners on antibodies; it does not save money in the long term!

(G) *Rinse after 1° antibody* – After 1° antibody incubation, all of the antibody must be diluted and washed away. Each rinse will be in 500 μl of rinse buffer and a total of six rinses, each done for 10 min with agitation. To correctly perform rinses, the key is not to totally remove all solution from the sample, which can cause drying of the sample. Rather, keep the number of rinses high. With a sufficient number of planned rinses, leaving 10–20% of the liquid at each rinse will allow removal of all of the previous incubation solution. Many labs are using as much as ten rinses to remove all of the antibody from the previous step

(H) *Incubate 2° antibody* – There are ten samples for 2° antibody (nine with 1° antibody and one without 1° antibody). With 250 μl per sample, the goat anti-mouse 488 fluorophore dilution is 1:1000. That will require 2.5 ml buffer plus 10% = 2.75 ml buffer and 2.8 μl of stock 1° antibody. The incubation with the 2° antibody is done for 1 h at room temperature.

(I) *Rinse after 2° antibody* – After 2° antibody incubations, a total of four rinses will be done with two different buffer solutions. First, rinse twice with 500 μl of rinse buffer for 5 min with agitation. Next, rinse twice with 500 μl of PBS alone for 5 min with agitation. The last buffer rinses are needed alone because some mounting media interact with proteins in the rinse buffer forming precipitates.

(J) *Mount coverslip – Use mounting medium without antifade reagents in the mounting medium.* Remove the last rinse of buffer and add a drop or two of mount medium to each well containing sections. The mounting medium rinse helps prevent mounting medium pulling away after fully drying. Before adding a coverslip, all of the hydrophobic material (e.g., PapPen) used to make the wells around the tissue sections must be removed. With the mounting medium in the wells, use a single-edge razor blade to scrape off all of the material that formed the wells. Tilt the slide on a paper towel to drain mounting medium and place several drops of fresh mounting medium on the sections. Lower a clear glass-labeled microscope slide. Do not attempt to remove mounting medium of the noncell surface of the coverslip facing up until the medium has hardened in 24 h.

Interrupt the procedure. Frequently, the entire procedure cannot be completed in 1 day, so plan for optimal times to stop the procedure, such as after fixation (store in refrigerator); after embedding in 20% sucrose (store in refrigerator or cold room); after freezing (store in –70°C freezer); after cutting and mounting sections on slides (store in refrigerator or cold room); and within the immunocytochemistry procedure at the last rinse after an antibody incubation (store in refrigerator or cold room). A great way to manage the procedure is to plan for more than 1 day and to perform an overnight primary antibody incubation.

(K) *Examine in microscope* – On the confocal microscope, configure the software and hardware for the fluorophore using a band pass filter for examining the 488 nm fluorophore (explanation in Chapter 13, Microscopy).

(L) *Evaluate results* – Initially, look at an experimental to determine whether the 2° antibody labeling is as expected. Next, examine the 2° antibody control section to confirm no labeling. Finally, determine that there are no other labeling problems (e.g., background labeling), and that the level of labeling is acceptable. If there are problems with these conditions, then repeat the experiment with corrective measures (Chapter 15, Trouble shooting). Otherwise proceed to evaluate the results of the experiment and determine if the results meet your scientific expectations.

Chapter 11
Multiple Antibodies Different Species

Keywords Immunohistochemistry · Antibody labeling · Fluorescence microscopy · Fluorescent immunocytochemistry · Fluorescent immunohistochemistry · Indirect immunocytochemistry · Immunostaining

Contents

Introduction 111
Combining Two 1° Antibody Incubations 112
Experimental Design Chart 112
Designing 2° Antibody Controls 113
Rules for Multiple Label Experiments 113
Complete Final Procedure 115
    (D) Block and Permeabilize 116
    (E) Rinse after Block and Permeabilize 116
    (F) 1° Antibodies 117
    (G) Rinse After 1° Antibody 117
    (H) 2° Antibody 117
    (I) Rinse After 2° Antibody 117

Introduction

The real power of immunocytochemistry is its utility to determine whether several different proteins are colocalized under specific experimental conditions. Therefore, most immunocytochemistry experiments use multiple 1° antibodies. Experiments with four fluorophores can be done with considerable planning.

There are two general strategies to use multiple 1° antibodies: (1) *1° antibodies made in different species* and (2) *1° antibodies made in the same species*. Initially, it might appear that this difference is not worth separate chapters; however, the important issue is how to handle the 2° antibodies. For experiments with 1° antibodies in different species, such as mouse and rabbit, 2° antibodies are made against the

IgG of either mouse or rabbit with little chance of binding the wrong antibody. In the case of experiments with 1° antibodies made in the same species, such as rabbit, both 2° antibodies are made against the IgG of rabbit and can bind to either 1° antibody. This chapter will describe the procedure for 1° antibodies in different species.

The experimental design presented in this chapter will combine the two single 1° antibody procedures into one procedure. Begin with determining the procedure for each 1° antibody separately (Chapter 10 Single 1° Antibodies). Then decide how to combine two 1° antibody incubations, design 2° antibody controls, and complete the final procedure.

Combining Two 1° Antibody Incubations

This chapter has an example of mouse anti-Ag A detected by goat anti-mouse 488 fluorophore and the rabbit anti-Ag B is detected by goat anti-rabbit 455 fluorophore. The antigen Ag A and antigen Ag B are on different proteins (Fig. 11.1a). The 1° antibodies were generated in different species of animals, mouse anti-Ag A and rabbit anti-Ag B (Fig. 11.1b). The 2° antibodies are goat anti-mouse 488 fluorophore and goat anti-rabbit 555 fluorophore (Fig. 11.1c). *Note that the 1° antibodies are made in the different species (mouse and rabbit) and the 2° antibodies are both made in goat but not mouse or rabbit, which removes the possibility that either of the 2° antibodies will react with each other or with the incorrect 1° antibody.* By eliminating the potential reactivity between the 2° antibodies, the use of 1° antibodies from different species is a powerful quick method. Finally, the fluorescent label used for each 2° antibody is unique. With 488 fluorophore and 555 fluorophore (Fig. 11.1c), separate identification of each 1° antibody is possible. With these conditions met, the antibodies for the 1° and 2° incubations can be combined into a single incubation.

Experimental Design Chart

Table 11.1 summarizes the controls for the experiments with 1° antibodies to Ag A and Ag B. Note that the antibody dilutions for the 1° and 2° antibodies were determined in the single antibody experiments that preceded this multiple antibody experiment. The first control uses the standard conditions with both 1° antibodies showing the expected normal labeling. The second control is No First 1° Antibody control, which eliminates the Ag A 1° antibody with the 488 fluorophore labeling. The Ag B labeling seen with 555 fluorophore should look normal. The third control, No Second 1° Antibody control eliminates the Ag B 1° antibody 555 fluorophore labeling. The Ag A labeling seen with 488 fluorophore should look normal. Finally, the fourth control, No First or Second 1° Antibody, eliminates both 1° antibodies and must eliminate all of the labeling with both 488 fluorophore and 555 fluorophore. In some cases, the 2° antibody controls will show unexpected labeling, which indicates that the 2° antibodies are binding to something other than the

Fig. 11.1 Multiple 1° antibodies different species. (**a**) Antigens Ag A and Ag B before incubations with antibodies. (**b**) The 1° antibody mouse anti-Ag A binds to Ag A and 1° antibody rabbit anti-Ag B binds to Ag B. (**c**) The 2° antibody goat anti-mouse 488 fluorophore binds to the mouse anti-Ag A and 2° antibody goat anti-rabbit 555 fluorophore binds to the rabbit anti-Ag B. The antibodies were combined in a single incubation for the 1° and the 2° antibodies

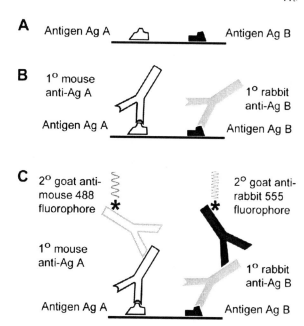

expected 1° antibody. To examine these situations, follow the trouble shooting steps in Chapter 15.

Designing 2° Antibody Controls

In previous experiments with single 1° antibodies (Chapter 10), the 2° antibody control showed that 2° antibody bound correctly and that its concentration was appropriate. With multiple 1° antibody experiments, the possibility exists that the 2° antibodies will bind unexpectedly in unforeseen ways. In the controls with multiple antibodies, an additional set of control conditions must be done each time the experiment is performed to insure that only the correct pair of 1° and 2° antibodies will bind.

The first set of controls are no 1° for each of the 1° antibodies (Table 11.2). This should show that the labeling for the 1° antibody was omitted and is not seen. The second control is to use neither of the 1° antibodies and eliminate all of the labeling.

Rules for Multiple Label Experiments

The important rules listed here summarize ideas from this chapter and the preceding chapters.

Table 11.1 Experimental design chart

Experimental design chart

Indirect immunocytochemistry

Category	Parameter	Conditions	Antigen No. 1	Antigen No. 2
(1) Sample				
	Source and tissue	Rat kidney		
	Fixative	4% paraformaldehyde		
	Fix application method	Perfused		
	Embedding	20% sucrose over night and isopentane freezing		
	Sectioning	Cryostat		
	Incubation chambers/size	Microscope slide 100 ml per area		
(2) 1° Antibody				
	Antigen		Ag A	Ag B
	Source of 1° antibody		Abs are Us No. 123	Ab are Us No. 321
	Species of 1° antibody		Mouse anti-Ag A	Rabbit anti-Ag B
	Dilution 1° antibody		Not known	1:1000
	Mixing 1° antibody		Not known	1 µl/ml
(3) 2° Antibody				
	Species		Goat anti-mouse	Goat anti-rabbit
	Source of 2° antibody		Abs are Us No. 987	Abs are Us No. 787
	Fluorophore		Fluo 488	Fluo 555
	Excitation wavelength		488 nm	546 nm
	Emission wavelength		505 nm	570 nm
	Dilution 2° antibody		Not known	1:1000
	Mixing 2° antibody		Not known	1 µl/ml
(4) Incubation solutions				
	Buffer	PBS pH 7.2		
	Block serum species	5% normal goat and 1 mg/ml BSA		
	Detergent	0.3% Triton		
(5) Controls				
	1° antibody controls		Done immunoblot product lit.	Done immunoblot product lit.
	2° antibody controls		Need to be done	Need to be done
	Label controls		Done no fluorescence	Done no fluorescence
(6) Microscope				
	Location	Confocal in core microscope facility		
	Fluorescent filters		488 nm band pass	546 long pass
	Lasers		Argon	Red HeNe

- Choose 1° antibodies from different species so that antibody incubations can be combined in a single step.
- Choose blocking serum from the same single species as the species of the 2° antibodies.
- Use third-generation fluorochromes for best fluorescence.
- Before trying multiple labels, optimize each fluorophore in a single-label experiment.

Table 11.2 Controls for indirect multiple antibodies different species

Controls for indirect multiple antibodies different species

Control type	Conditions	First 1°	Secondary 1°
Experimental	1° Antibody	Ag A	Ag B
	2° Antibody	Goat anti-mouse 488 fluorophore	Goat anti-mouse Alexa 543 fluorophore
	Labeling observed	Normal (green)	Normal (red)
No first 1° antibody	1° Antibody	None	Ag B
	2° Antibody	Goat anti-mouse 488 fluorophore	Goat anti-mouse Alexa 543 fluorophore
	Labeling observed	No labeling	Normal (red)
No second 1° antibody	1° Antibody	Ag A	None
	2° Antibody	Goat anti-mouse 488 fluorophore	Goat anti-mouse Alexa 543 fluorophore
	Labeling observed	Normal (green)	No labeling
No first or second 1° antibody	1° Antibody	None	None
	2° Antibody	Goat anti-mouse 488 fluorophore	Goat anti-mouse Alexa 543 fluorophore
	Labeling observed	No labeling	No labeling

- Put the most intense fluorochrome at the highest wavelength fluorochrome.
- Choose fluorochromes so excitations (and therefore emissions) are separated by 80–100 nm.

Complete Final Procedure

Steps in a two 1° antibody different species indirect immunocytochemistry experiment.

(A) Prepare animal
(B) Fix tissue

(C) Embed, section, and mount tissue
(D) Block and permeabilize
(E) Rinse after block and permeabilize
(F) Incubate 1° antibody
(G) Rinse after 1° antibody
(H) Incubate 2° antibody
 (I) Rinse after 2° antibody
(J) Mount coverslip
(K) Examine in microscope
(L) Evaluate results

Steps A–D and J–L that were covered in Chapter 10 for the single antibody procedure are the same and will not be repeated here. The discussion below will focus on the steps necessary for antibody incubations from D through I. Because both 1° antibodies and 2° antibodies can be combined in single incubation steps, the number of steps and the amount of reagents will be similar to those in the single 1° antibody procedure.

(D) Block and Permeabilize

Because both 2° antibodies are produced in goat, the reagents in this step are identical to those used for the single 1° antibody. The blocking and permeabilization solution contains 10 mg/ml BSA and 5.0% normal goat serum in PBS to block nonspecific antibody binding. Permeabilize with 0.3% Triton and 0.02% sodium azide to prevent bacterial growth during incubations. Incubation is for 1 h.

(E) Rinse after Block and Permeabilize

The rinse buffer solution contains 10 mg/ml BSA and 5.0% normal goat serum in PBS (no detergent); use this for all rinses. Mix only as much as needed for the current experiment. To prepare this solution, multiply the number of samples times the volume of each rinse, times the total number of rinses for the entire experiment. For this example experiment, we have 17 experimental samples and 3 controls samples for 20 total samples. For the rinse buffer, use 500 μl for each rinse and a total of 14 rinses (two rinses after block and permeabilize; six rinses after 1° antibody; six rinses after 2° antibody). A total of 20 samples × 14 total rinses × 0.5 ml per rinse = 140 ml of rinse buffer plus 5% excess (7 ml) is 147 ml. The excess amount is a safety factor, in case of a spill or other unexpected event.

The rinses after the block and permeabilize step is to remove the Triton so that it will only act for a specific period of time. Two rinses are sufficient to dilute the Triton. Rinse for 5 min with the same agitation used during the block and permeabilize step.

(F) 1° Antibodies

The two 1° antibodies are made in different species and can be combined in one incubation solution. The mouse anti-Ag A at 1:500 and the rabbit anti-Ag B at 1:1000 are mixed in a single solution. For the 17 samples in this experiment, but not the three 2° control conditions, use 250 μl per sample, plan for 4.25 ml plus 10% excess or 4.7 ml of the rinse buffer. The mouse anti-Ag A at 1:500 will be 9.4 μl of stock and the rabbit anti-Ag B at 1:1000 will be 4.7 μl. The incubation with the 1° antibody is done for 6 h at room temperature.

(G) Rinse After 1° Antibody

After incubation with combined 1° antibodies, all of the antibody must be rinsed away. Each rinse will be in 500 μl of rinse buffer and a total of four rinses for 5 min with agitation.

During all rinses, as stated previously, the key is to leave solution on the sample, which will prevent drying of the sample and thus prevent drying of the section. After each rinse, leave 10–20% of the liquid on the sample and do not remove all of the solution. Drying of the sample will dislodge cells on the coverslip and permanently attach antibodies to the cells which increases background. A minimum of six 5–10 min rinses are recommended. Many labs use as much as 10 rinses to remove all of the antibody from the previous step

(H) 2° Antibody

As with the 1° antibodies, the 2° antibodies can be combined into one incubation solution, with goat anti-rabbit 555 fluorophore at 1:1000 and the goat anti-mouse 488 fluorophore at 1:1000. Both of these antibodies are diluted in a single buffer solution. For the 20 samples in this experiment at 250 μl per sample, plan for 5 ml plus 10% excess or 5.5 ml of the rinse buffer. The goat anti-mouse 488 fluorophore at 1:1000 will be 5.5 μl of stock and the goat anti-rabbit 555 fluorophore at 1:1000 will be 5.5 μl. The incubation with the 1° antibody is done for 1 h on a rocker at room temperature.

(I) Rinse After 2° Antibody

After completing 2° antibody incubations, all of the antibody must be washed away. In addition, some mounting media can interact with proteins in the rinse buffer forming precipitates. Plan a total of six rinses done with two different solutions. First, four rinses will be in 500 μl of rinse buffer for 5 min with agitation. Next, two rinses of 500 μl of PBS for 5 min with agitation.

Chapter 12
Multiple Antibodies from the Same Species

Keywords Immunohistochemistry · Antibody labeling · Fluorescence microscopy · Fluorescent immunocytohemistry · Fluorescent immunohisto-chemistry · Indirect immunocytochemistry · Immunostaining

Contents

Introduction 120
Combine Two 1° Antibodies from the Same Species
    with Block-Between Method 120
Experimental Design Chart for Block-Between Method 122
Design the 2° Antibody Control for the Same Species
    with Block-Between 124
Final Procedure for Two 1° Antibody Same Species
    with Block-Between 127
  (A) Prepare Cell Culture 127
  (B) Fix Culture 127
  (C) Block and Permeabilize 127
  (D) Rinse After Block and Permeabilize 128
  (E) Incubate First 1° Antibody 128
  (F) Rinse After First 1° Antibody 128
  (G) Incubate First 2° Antibody 128
  (H) Rinse After First 2° Antibody 128
  (I) Block Antibodies in First Set 128
  (J) Incubate Second 1° Antibody 128
  (K) Rinse After Second 1° Antibody 129
  (L) Incubate Second 2° Antibody 129
  (M) Rinse After Second 2° Antibody 129
  (N) Mount Coverslip 129
  (O) Examine in Microscope 129
  (P) Evaluate Results 129

R.W. Burry, *Immunocytochemistry*, DOI 10.1007/978-1-4419-1304-3_12,
© Springer Science+Business Media, LLC 2010

Combine Two 1° Antibodies from the Same
    Species with Zenon   130
Experimental Design Chart for the Same Species with Zenon   130
Design the Antibody Control for the Same Species with Zenon   133
Final Procedure for Two 1° Antibody from the Same Species with Zenon   135
    (A) Prepare Cell Culture   135
    (B) Fix Culture   135
    (C) Block and Permeabilize   136
    (D) Rinse after Block and Permeabilize   136
    (E) Prepare the Zenon Reagents   136
    (F) Incubate with Labeled Antibody(ies)   136
    (G) Rinse After Antibody Incubation   136
    (H) Fix with 4% Paraformaldehyde   136
    (I) Rinse after Antibody Incubation   137
    (J) Mount Coverslip   137
    (K) Examine in Microscope   137
    (L) Evaluate Results   137

Introduction

Using 1° antibodies made in the different species is the easiest way to localize multiple proteins (Chapter 11). However, it is not possible to get all 1° antibodies needed from a different species. This is especially true because there are so many mouse monoclonal antibodies available. Eventually, an experiment will need two mouse 1° antibodies or two rabbit 1° antibodies. This chapter presents the concept of combining *multiple 1° antibodies made in the same species* of animals. Two different approaches include *block-between method* and labeled Fab procedure (Lewis et al., 1993) that is available as the commercial product, *Zenon*® (Molecular Probes/Invitrogen).

Combine Two 1° Antibodies from the Same Species with Block-Between Method

When an experiment uses multiple 1° antibodies derived from the same species, perform the incubations for the first 1° and 2° antibodies, block the remaining antibody sites, and then perform the incubations for the second 1° and 2° antibodies. This procedure consists of a series of two single 1° antibody indirect immunocytochemistry experiments with extensive blocking between them. The key element is the blocking steps between. In this example, cultures are incubated with the first 1° antibody set, mouse anti-Ag A and 2° antibody goat anti-mouse labeled with 488 fluorophore (Fig. 12.1a), followed by steps that block the remaining antibody sites (Fig. 12.1b, c). The incubations with the second 1° antibody set are with mouse anti-Ag B antibody and 2° antibody goat anti-mouse

Fig. 12.1 Multiple 1° antibodies for the same species with block-between. After the incubations with the 1° and 2° antibodies for antigen Ag A are completed, special blocking steps were completed, followed by the incubations with 1° and 2° antibodies for antigen Ag B. (a) Antigen Ag A is bound by mouse anti-Ag A 1° antibody and then bound by a goat anti-mouse 2° with a 488 fluorophore. No incubations are done for antigen Ag B. (b) The first step is to block the antibodies with normal mouse serum containing IgG. The IgG binds any available anti-mouse site on the 2° antibody. (c) The next blocking step is to block the species-specific sites on the constant region of any mouse antibody. Anti-mouse Fab fragments are incubated and bind to both the 1° antibody and to the normal serum IgG used in the first blocking step. (d) With all of the sites blocked, the next set of antibodies is incubated with no chance of binding to the first set of antibodies. Antigen Ag B is bound by mouse anti-Ag B 1° antibody and then bound by a goat anti-mouse 2° with a 555 fluorophore

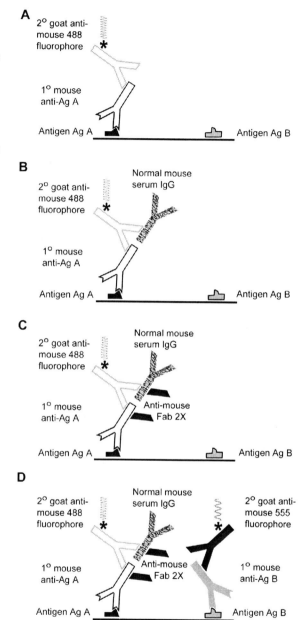

with 555 fluorophore (Fig. 12.1d). This procedure was originally described by Lewis et al. (1993).

After the first 1° and 2° antibody incubation, the block-between consists of two steps. The first blocking step uses normal mouse serum to block any unbound mouse Fab sites from anti-mouse 2° antibody. The first 2° antibody, goat anti-mouse 488 fluorophore, has unbound sites (Fig. 12.1a), which are *blocked by the use of normal mouse serum containing nonimmune IgG molecules* (Fig. 12.1b, normal mouse serum IgG). These normal mouse IgG molecules bind to the available anti-mouse Fab sites so that the next mouse 1° antibody incubation will not bind to them. Rinses are used to wash away all unbound nonimmune IgG.

The second blocking step blocks any mouse species-specific sites on the mouse antibodies. There are two sources of mouse IgG that can have exposed mouse species-specific sites, which could incorrectly bind the second 2°. The potential sources are unbound sites on the mouse anti-Ag A 1° antibody and the nonimmune IgG molecules used to block in the previous step (Fig. 12.1b). To *block these mouse-specific sites, use an anti-mouse Fab fragment* (Fig. 12.1c, black Fab). The Fab fragment is a papain-digested IgG that has only Fab ends (Chapter 2, Antibodies). The species of the animal used to generate this fragment does not matter, because the species-specific end, Fc, has been removed. Incubation with the anti-mouse Fab will block all of the species-specific sites on both the first 1° antibody and the blocking normal mouse IgG (Fig. 12.1c, anti-mouse Fab 2X), which will prevent binding of the next 2° antibody. Rinses are required to remove the excess Fab. These two blocking steps must be done in the order with normal mouse serum first and anti-mouse Fab fragments second. With these two blocking incubations complete, the second 1° antibody, mouse anti-Ag B, binds to its antigen. The second 2° antibody is goat anti-mouse 555 fluorophore and binds only to the second 1° antibody, mouse anti-Ag B (Fig. 12.1d).

The great power of this method is that it allows 1° antibodies generated in the same species of animal to be used together, even three or four at a time. The downside of this method is that it has many steps, many rinses, takes time, and there are many controls that must be performed. But when antibodies made in mouse are the only antibodies available, it is a good choice.

Experimental Design Chart for Block-Between Method

A new Experimental Design Chart must be used for this Indirect Immunocytochemistry Block-Between (Table 12.1), because it differs from the chart used previously. Here, the experiment uses cultured cells on coverslips and two 1° antibodies made in mouse, mouse anti-Ag A and mouse anti-Ag B. A new row on the Chart is needed, 4. Block-between, to list two new reagents used to block-between the two 1° antibodies. Normal mouse serum is used at 1:20, which is sufficient to block in most cases. The anti-mouse Fab molecule is used at 20 μg/ml. Incubations for each of these steps is for 1 h with six rinses after each incubation. Note these blocking incubations must be performed in the order of normal mouse serum first and anti-mouse Fab second.

Table 12.1 Experimental design chart

Experimental design chart
Indirect immunocytochemistry block-between

Category	Parameter	Conditions	Antigen No. 1	Antigen No. 2
(1) Sample				
	Source and tissue	PC12 cell line		
	Fixative	4% paraformaldehyde		
	Fix application method	Cultures		
	Embedding	None		
	Sectioning	None		
	Incubation chambers/size	12 mm coverslip in 24-well plate		
(2) 1° Antibody				
	Antigen		Ag A	Ag B
	Source of 1° antibody		Abs are Us No. 123	Ab are Us No. 321
	Species of 1° antibody		Mouse anti-Ag A	Mouse anti-Ag B
	Dilution 1° antibody		1:800	1:200
	Mixing 1° antibody		1.25 µl/ml	5 µl/ml
(3) 2° Antibody				
	Species		Goat anti-mouse	Goat anti-mouse
	Source of 2° antibody		Abs are Us No. 987	Abs are Us No. 787
	Fluorophore		488 fluorophore	555 fluorophore
	Excitation wavelength		488 nm	546 nm
	Emission wavelength		505 nm	570 nm
	Dilution 2° antibody		1:1000	1:1000
	Mixing 2° antibody		1 µl/ml	1 µl/ml
(4) Block-between				
	Dilution normal 1° species serum		1:20	
	Dilution of anti-species 1° Fab		20 µg/ml	
(5) Incubation solutions				
	Buffer	PBS pH 7.2		
	Block serum species	5% normal goat and 1 mg/ml BSA		
	Detergent	0.3% Triton		
(6) Controls				
	1° antibody controls		Done immunoblot product lit.	Done immunoblot product lit.
	2° antibody controls		Need to be done	Need to be done
	Label controls		Done no fluorescence	Done no fluorescence
(7) Microscope				
	Location	Wide field fluorescence microscope in lab		
	Fluorescent filters		488 nm band pass	546 long pass
	Lasers		NA	NA

Design the 2° Antibody Control for the Same Species with Block-Between

Plan to add controls that confirm successful blocking steps between two sets of antibodies (Table 12.2). Because of the sequential addition of antibodies, the controls are different from other experiments with the indirect method of immunocytochemistry. The first 1° antibody is not eliminated because there are no competing antibodies for the first 1° antibody. Also, the no 1° antibody control for the first 1° antibody was done previously when the Dilution Matrix showed it was bound specifically by the 2° antibody. The controls here test the potential binding of the second 1° antibody and second 2° antibody to the first set of antibodies.

The first control removes the second 1° antibody (Table 12.2), which will show that the second 2° antibody is binding only to the second 1° antibody. The second control reverses the order of the sets of antibodies, so that anti-Ag B is now first and anti-Ag A is second. This will confirm that the second 2° antibody is binding to the correct 1° antibody.

To see the importance of the blocking steps let us look at examples where the steps in the block-between were omitted. Incorrect binding of the second 2° antibody will occur if the normal mouse serum does not block all of the first 2° antibodies. The first set of incubations with mouse anti-Ag A is followed by a goat anti-mouse 488 fluorophore antibody (Fig. 12.2a). When this first 2° antibody is not blocked, then the second 1° mouse anti-Ag B will bind (Fig. 12.2b, solid gray antibody). When the second 2° antibody, goat anti-mouse 555 fluorophore, is used it binds to 1° antibody in both places and it will label correctly the Ag B antigen and incorrectly for the Ag A antigen. The use of the no second 1° antibody control will show this incorrect binding.

Table 12.2 Controls for indirect immunocytochemistry multiple antibodies with block between

Control type	Conditions	First 1°	Secondary 1°
Experimental	1° Antibody	Ag A	Ag B
	2° Antibody	Goat anti-mouse 488 fluorophore	Goat anti-mouse 543 fluorophore
	Labeling observed	Normal (green)	Normal (red)
No Second 1° antibody	1° Antibody	Ag A	None
	2° Antibody	Goat anti-mouse 488 fluorophore	Goat anti-mouse 543 fluorophore
	Labeling observed	No labeling	Normal
Reverse order	1° Antibody	Ag B	Ag A
	2° Antibody	Goat anti-mouse 543 fluorophore	Goat anti-mouse 488 fluorophore
	Labeling observed	Normal (red)	Normal (green)

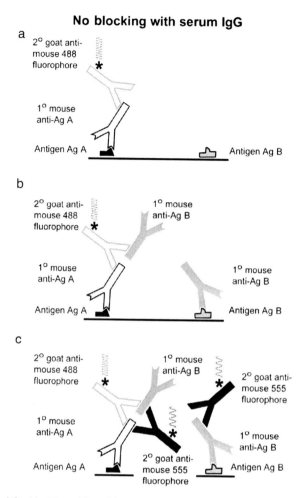

Fig. 12.2 Need for blocking with multiple 1° antibodies block-between. If the blocking step is not done or is not complete, the second 2° antibody will bind both 1° antibodies. (**a**) The first 1° antibody is a mouse anti-Ag A; it is bound by the first 2° antibody a goat anti-mouse 488 fluorophore. (**b**) With no blocking steps, the second 1° antibody mouse anti-Ag B binds to the antigen Ag B and also to the first 2° the goat anti-mouse 488 fluorophore. (**c**) The second 2° antibody, goat anti-mouse 555 fluorophore, binds to the mouse anti-Ag B and labels the sites for both antigens

Another potential problem occurs if the anti-mouse Fab molecules do not block the mouse species-specific sites. Incubation with the mouse anti-Ag A and anti-mouse 488 fluorophore (Fig. 12.3a) is then blocked with normal mouse serum IgG (Fig. 12.3b, stippled gray). If there is insufficient anti-mouse Fab to block the mouse antibodies from the first set of incubations, then the second 2° antibody will bind to the first set of antibodies. Specifically, the second 1° antibody, mouse anti-Ag B, binds to the Ag B antigen (Fig. 12.3c). Then the lack of anti-mouse Fab fragments

Fig. 12.3 Blocking multiple 1° antibodies block between; If the second blocking step with the anti-mouse Fab is not done or is not complete, the second 2° antibody will bind both 1° antibodies. (a) The first 1° antibody is a mouse anti-Ag A, which bound by the first 2° antibody, a goat anti-mouse 488 fluorophore as before. (b) The blocking with the normal mouse serum IgG binds as expected to the first 2° antibody, goat anti-mouse 488 fluorophore. (c) Incubation with the second 1° antibody gives normal binding to antigen Ag B. (d) Incubation with the second 1° antibody mouse anti-Ag B has binding to both antigens bound by mouse antibodies

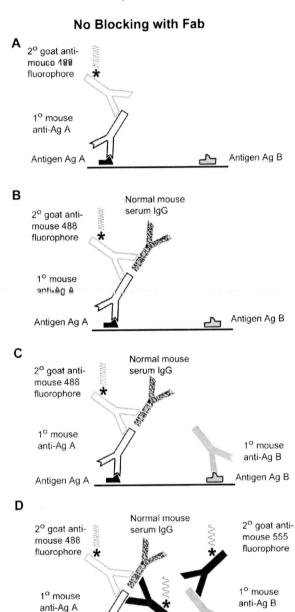

will allow the second 2° antibody, goat anti-mouse 555 fluorophore, to bind (Fig. 12.3d). Thus the binding of the second 2° antibody to the first set of antibodies results in the incorrect 555 fluorophore label for both anti-Ag A and anti-Ag B.

Final Procedure for Two 1° Antibody Same Species with Block-Between

Steps in a two 1° antibody same species block-between experiment.

(A) Prepare cell culture
(B) Fix culture
(C) Block and permeabilize
(D) Rinse after block and permeabilize
(E) Incubate first 1° antibody
(F) Rinse after first 1° antibody
(G) Incubate first 2° antibody
(H) Rinse after first 2° antibody
 (I) Block antibodies in first set
 (J) Incubate second 1° antibody
(K) Rinse after second 1° antibody
(L) Incubate second 2° antibody
(M) Rinse after second 2° antibody
(N) Mount coverslip
(O) Examine in microscope
(P) Evaluate results

(A) Prepare Cell Culture

For this example experiment, cell cultures were prepared on 12 mm glass coverslips in a 24-well plastic culture dish.

(B) Fix Culture

The culture medium is removed and the cultures are rinsed twice in PBS. The PBS is removed and the fixative, 4% paraformaldehyde in PBS, is added for 30 min.

(C) Block and Permeabilize

Blocking and permeabilization solution contains 10 mg/ml BSA, 5.0 % normal goat serum to block nonspecific antibody binding and 0.02% sodium azide is used to prevent bacterial growth during incubations all in PBS. Perform permeabilization with 0.2% Triton. Incubate for 1 h.

(D) Rinse After Block and Permeabilize

Use rinse buffer solution called PBS plus contains 10 mg/ml BSA, 5.0% normal goat serum, and 0.02% sodium azide in PBS (no detergent).

(E) Incubate First 1° Antibody

In this example experiment, the mouse anti-Ag A 1° antibody has been used in the lab and its dilutions has been determined to be 1:800.

(F) Rinse After First 1° Antibody

After 1° antibody incubation, all of the antibody must be diluted and washed away. Each rinse will be in 500 μl of PBS plus rinse buffer and a total of six rinses, each done for 10 min with agitation. For cultures, allowing a sample to drying will tear cells off the coverslip and drying will permanently attach antibodies to the remaining cells increasing background. Rinse a minimum of five times for 5–10 min each.

(G) Incubate First 2° Antibody

The 2° antibody as found in the Dilution Matrix is goat anti-mouse 488 fluorophore; dilution is 1:1000

(H) Rinse After First 2° Antibody

After 2° antibody incubation, all of the antibody must be diluted and washed away. Each PBS plus rinse will be in 500 μl of rinse buffer and a total of six rinses, each done for 10 min with agitation.

(I) Block Antibodies in First Set

Incubate with 1:20 normal mouse serum (Jackson ImmunoResearch 015-000-120) for 1 h, followed by six rinses with PBS plus 5 min each. The second blocking step is incubation with goat anti-mouse Fab (Jackson ImmunoResearch 115-007-003) 20 μg/ml or 14 μl/ml for 1 h. This blocking step is ended with six rinses with PBS plus 5 min each.

(J) Incubate Second 1° Antibody

The second 1° antibody is mouse anti-Ag; its dilution has been determined to be 1:200.

(K) Rinse After Second 1° Antibody

After 1° antibody incubation, each rinse will be in 500 μl of PBS plus rinse buffer and a total of six rinses, each done for 10 min with agitation.

(L) Incubate Second 2° Antibody

The second 2° antibody as found in the Dilution Matrix is goat anti-mouse 555 fluorophore; dilution is 1:1000

(M) Rinse After Second 2° Antibody

After 2° antibody incubation, each rinse will be in 500 μl of PBS plus rinse buffer and a total of six rinses, each done for 10 min with agitation.

(N) Mount Coverslip

To remove the coverslips from the wells of the 24-well plate, use Dumont 3C forceps with tape on one side and a long syringe needle with bent at a right angle near the end. Use the bent syringe needle to lift the edge of the coverslip and to hold it off the substrate. Next open the forceps and grab the edge of the coverslip. Always hold the coverslip in 3C forceps so that the cells are on the side with the tape label you have previously added. Put a drop of mount medium on Parafilm, which is taped to the countertop. Place the coverslip tissue side down to float on the drop of mounting medium for 5 min. Remove the coverslip, drain excess mounting medium, and place a small drop of mounting medium on a clear glass-labeled microscope slide. Do not attempt to remove mounting medium of the noncell surface of the coverslip facing up until the medium has hardened in 24 h.

(O) Examine in Microscope

With wide field fluorescence microscope, examine the cultures for fluorescence with a 488 band pass filter set and with a 546 long pass filter set (explanation in Chapter 13, Microscopy and Images). With the CCD camera configure the software and hardware for capturing the images.

(P) Evaluate Results

Initially, look at an experimental to determine whether the 2° antibody labeling is as expected. Next, examine the 2° antibody control section to confirm no labeling. Finally, determine that there are no other labeling problems (e.g., background labeling) and that the level of labeling is acceptable. If there are problems with these conditions, then repeat the experiment with corrective measures (Chapter 14, Troubleshooting). Otherwise, proceed to evaluate the results of the experiment and determine if the results meet your scientific expectations.

Combine Two 1° Antibodies from the Same Species with Zenon

The second approach is very different and involves labeling the 1° antibody before incubation with tissues. In effect, this makes the procedure a direct immunocyto-chemistry approach because no 2° antibody is used. *The method uses labeled Fab molecules from anti-species antibodies that bind to the Fc end of the 1° antibody.* The labeled Fab reagents needed for this method are available as a commercial kit called Zenon® (Molecular Probes/ Invitrogen).

To use this Zenon method, the individual 1° antibodies are labeled individually with fluorescent-labeled Fab before the immunocytochemistry incubation with cells. To do this, incubate with the 1° antibody, mouse anti-Ag A, with anti-mouse Fab 488 fluorophore reagent (Fig. 12.4a) and the labeled fluorophore, which binds to the 1° antibody (Fig. 12.4b). Excess labeled 488 fluorophore Fab reagent must be present in the incubation solution (Fig. 12.4b). The blocking reagent, normal mouse IgG contained in nonimmune mouse serum, binds the excess 488 fluorophore-labeled anti-mouse Fab (Fig. 12.4c). The mouse anti-Ag A labeled with Fab 488 fluo-rophore is available for use. The excess normal mouse IgG, either free or bound to anti-mouse Fab 488 fluorophore, will not bind to anything during the rest of the experiment.

In the next step, the mouse anti-Ag B is labeled with anti-mouse Fab 555 flu-orophore (Fig. 12.4d). For multiple label immunocytochemistry combine the two labeled 1° antibodies and use directly without purification. The incubation solution contains the two labeled 1° antibodies as well as labeled normal mouse IgG for each fluorophore and excess normal mouse IgG (Fig. 12.4e).

In planning the incubation with both labeled 1° antibodies, be aware the Fab binding to the 1° antibody is not permanent. For 1 h, the binding is saturated, but after several hours, labeled Fab begins to dissociate with only 75% of the 1° antibody labeled. Thus, labeled 1° antibody must be prepared just before use and follow the incubation fixed with paraformaldehyde to retain the label attached to the antibody.

Experimental Design Chart for the Same Species with Zenon

To plan an experiment, refer to the Experimental Design Chart for Indirect Immunocytochemistry Zenon with mouse anti-Ag A and a mouse anti-Ag B 1° antibodies (Table 12.3). In the chart Section 2. 1° *Antibodies* include the isotype of each mouse 1° antibodies. Use this information in selecting a Fab-labeling reagent for the 1° antibodies. Use chart Section 3. 1° *Antibody Labeling,* to help select the

Fig. 12.4 Multiple 1°
antibodies for the same
species with Zenon. 1°
antibodies are labeled
independently and then mixed
together for a single
incubation step. (**a**) In a tube,
the 1° antibody mouse
anti-Ag A is incubated and
labeled with anti-mouse Fab
labeled with 488 fluorophore.
(**b**) Excess anti-mouse Fab
labeled with 488 fluorophore
is used to bind all the possible
binding sites on the 1°
antibody mouse anti-Ag A.
(**c**) The excess anti-mouse
Fab labeled with 488
fluorophore is bound to
normal mouse serum IgG,
which is then added to the
tube. Now the 488
fluorophore-labeled 1°
antibody mouse anti-Ag A is
now ready to use. (**d**) In a
separate tube, 1° antibody
mouse anti-Ag B incubated
with anti-mouse Fab labeled
with 555 fluorophore
followed by normal mouse
serum IgG. (**e**) Both tubes are
mixed and added to sections
that contain antigens Ag A
and Ag B. The solution
contains normal mouse serum
IgG bound to 488
fluorophore, normal mouse
serum IgG bound 555
fluorophore, and unbound
normal mouse serum IgG,
none of which affect the
labeling seen with the labeled
1° antibodies

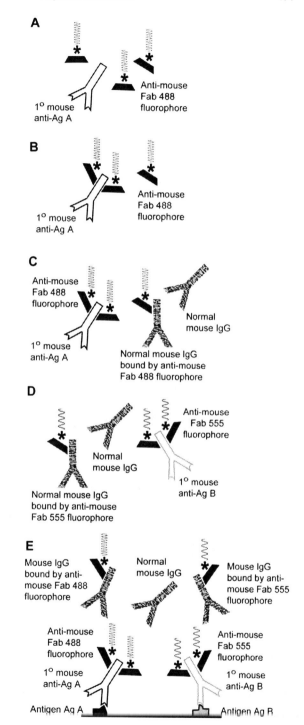

Table 12.3 Experimental design chart

Experimental design chart
Indirect immunocytochemistry Zenon®

Category	Parameter	Conditions	Antigen No. 1	Antigen No. 2
(1) Sample				
	Source and tissue	PC12 cell line		
	Fixative	4% Paraformaldehyde		
	Fix application method	Cultures		
	Embedding	None		
	Sectioning	None		
	Incubation chambers/size	12 mm coverslip in 24-well plate 300 ml per well		
(2) 1° Antibody				
	Antigen		Ag A	Ag B
	Source of 1° antibody		Abs are Us No. 123	Ab are Us No. 321
	Species of 1° antibody		Mouse anti-Ag A	Rabbit anti-Ag B
	1° Antibody subclass		IgG2a	IgG2b
	Dilution 1° antibody X 2		1:800 X 2 = 1:400	1:200 X 2 = 1:100
	Mixing antibody		2.5 µl/µg	10 µl/µg
(3) 1° Antibody labeling				
	Source of Fab		Abs are Us No. 987	Abs are Us No. 787
	Fluorophore		488 fluorophore	555 fluorophore
	Excitation wavelength		488 nm	546 nm
	Emission wavelength		505 nm	570 nm
	Amount Fab zenon component A		20 µl/µg 1° Ab	20 µl/µg 1°Ab
	Amount normal mouse serum		5 µl/µg 1°Ab	5 µl/µg 1°Ab
(4) Incubation solutions				
	Buffer	PBS pH 7.2		
	Block serum species	5% normal goat and 1 mg/ml BSA		
	Detergent	0.3% Triton		
(5) Controls				
	1° antibody controls		Done immunoblot product lit.	Done immunoblot product lit.
	2° antibody controls		Need to be done	Need to be done
	Label controls		Done no fluorescence	Done no fluorescence
(6) Microscope				
	Location	Wide field fluorescence microscope in lab		
	Fluorescent filters		488 nm band pass	546 nm long pass
	Lasers		NA	NA

Fab-labeled reagents based on the species of the 1° antibody and the fluorophore needed.

The Fab-labeling reagent from molecular probes/invitrogen is called "Zenon mouse IgG labeling reagent" or "Component A." For immunocytochemistry, the molar dilution of the labeled Fab reagent to the 1° antibody is 6:1 by weight (labeled Fab reagent to 1° antibody). This ratio provides an excess of the labeled Fab reagent. For example, if 2 μg of 1° antibody is used, then 6 μg of labeled Fab reagent (3 moles of Fab per μg) will be needed. The Zenon reagent, labeled Fab reagent, comes as 0.2 μg/μl and 5 μl equals 1 μg labeled Fab. For 2 μg of 1° antibody, 20 μl Zenon mouse IgG labeling reagent is needed. The total reaction volume is not crucial but should be 20 μl or more.

Finally, to block the excess labeled Fab reagent molecules, add 10 μl of normal mouse serum containing IgG to the above solution for a second incubation. The amount of normal mouse serum by weight should be five times that of the 1° antibody. When 2 μg of 1° antibody is used then 10 μg of normal mouse serum is needed. Each individual 1° antibody must be incubated with the labeled Fab reagent and the blocking normal mouse serum before being combined into a single antibody incubation solution.

For this method it is recommended that the 1° antibody be used at twice the dilution used for standard indirect immunocytochemistry. If the 1° antibody dilution is 1:800, twice the dilution will be 1:400. To make 1 ml for the final use 2.5 μl of 1° antibody to mix with 20 μl of labeled Fab and incubated for 5 min with agitation. Then 10 μl of blocking normal mouse serum is added and incubated with agitation for 5 min. Calculating the dilution for the primary antibody needs to consider that enough 1° antibody is used for a final volume of 1 ml. For multiple 1° antibodies the total final volume should be 1 ml for all antibodies mixed together. If the concentration of the 1° antibody is not known, then do a dilution series with different amount of 1° antibody and the same amount of Zenon mouse IgG-labeling reagent and the blocking serum reagent.

The most common problems with this method are that little or no labeling is seen with the final labeled antibody rather than too much labeling. Correct this situation by adding more labeled Fab reagent to the 1° antibody and increase the ratio from 6:1 to 10:1. Changing the ratio will require that the control experiments be performed with antibody labeled with the higher ratio.

Design the Antibody Control for the Same Species with Zenon

Two antibody controls are needed for labeled Fab 1° antibody Zenon immunocytochemistry (Table 12.4). These controls will show that each 1° antibody is labeled correctly. To generate a control and additional 1° antibody labeling reaction is needed, where the blocking normal mouse serum is added to the labeled Fab reagent (Zenon reagent) before the 1° antibody. In this incubation, the labeled

Fab reagent will bind the normal mouse serum before the $1°$ antibody is added and no labeling of the $1°$ antibody should be found. If the single $1°$ antibody sample has any labeling then the blocking antibody (normal mouse serum) concentration is too low.

The first control is the No first $1°$ labeling (Table 12.4) with $1°$ antibody, mouse anti-Ag A, which should show that the normal mouse IgG blocks all the labeled Fab fragment binding. To perform this control, an additional labeling reaction is needed, in which the reagents are added in the reverse order. To the anti-mouse Fab 488 fluorophore, add the nonspecific IgG normal mouse serum first and then add the mouse anti-Ag A $1°$ antibody. Next, use mouse anti-Ag B labeled normally and mix the Ag A and the Ag B $1°$ antibody solutions. Following incubations of cells, there should be no labeling for the mouse Fab 488 fluorophore because it binds to the IgGs in the blocking serum and leave none to bind the mouse Ag A antigen.

Table 12.4 Controls for indirect multiple antibodies same species Zenon

Controls for indirect multiple antibodies same species Zenon

Control type	Conditions	First $1°$ mixing	Second $1°$ mixing
Experimental	$1°$ Antibody	Mouse anti-Ag A	Mouse anti Ag B
	First incubation	Anti-mouse 488 fluorophore	Anti-mouse 555 fluorophore
	Second incubation	Normal mouse serum	Normal mouse serum
	Labeling observed	Normal (green)	Normal (red)
No first $1°$ antibody	$1°$ antibody	Normal mouse serum	Mouse anti Ag B
	First incubation	Anti-mouse 488 fluorophore	Anti-mouse 555 fluorophore
	Second incubation	Mouse anti Ag A	Normal mouse serum
	Labeling observed	No labeling	Normal (red)
No second $1°$ antibody	$1°$ antibody	Mouse anti Ag A	Normal mouse serum
	First incubation	Anti-mouse 488 fluorophore	Anti-mouse 555 fluorophore
	Second incubation	Normal mouse serum	Mouse anti Ag B
	Labeling observed	Normal (green)	No labeling

After using this control step, no labeling should be seen but normal mouse anti-Ag B should be seen.

The second control is the No second 1° labeling (Table 12.4) with mouse anti-Ag B 1° antibody and should show that the normal mouse IgG blocks all the labeled Fab fragment binding. Perform this control using the pattern established above. To the anti-mouse Fab 555 fluorophore, add the nonspecific IgG normal mouse serum and then add the mouse anti-Ag B 1° antibody. Next, mix the anti-Ag A and the anti-Ag B 1° antibodies and incubate with the cells. Again, no labeling should be seen when this solution is incubated with samples. This control must be done to confirm that the amount of blocking is still sufficient, especially if the ratio of labeled Fab to 1° antibody has been altered.

Final Procedure for Two 1° Antibody from the Same Species with Zenon

Steps in a two 1° antibody same species Zenon experiment.

(A) Prepare cell culture
(B) Fix culture
(C) Block and permeabilize
(D) Rinse after block and permeabilize
(E) Prepare the Zenon reagents
(F) Incubate with labeled antibody(ies)
(G) Rinse after antibody incubation
(H) Fix with 4% paraformaldehyde
 (I) Rinse with PBS plus
 (J) Mount coverslip
(K) Examine in microscope
(L) Evaluate results

(A) Prepare Cell Culture

For this example experiment, cell cultures were prepared on 12 mm glass coverslips in a 24-well plastic culture dish.

(B) Fix Culture

The culture medium is removed and the cultures are rinsed twice in PBS. The PBS is removed and the fixative, 4% paraformaldehyde in PBS, is added for 30 min.

(C) Block and Permeabilize

Blocking and permeabilization solution contains 10 mg/ml BSA, 5.0 % normal goat serum to block nonspecific antibody binding, and 0.02% sodium azide is used to prevent bacterial growth during incubations all in PBS. Perform permeabilization with 0.2% Triton. Incubate for 1 h.

(D) Rinse after Block and Permeabilize

Use rinse buffer solution called PBS plus that contains 10 mg/ml BSA, 5.0 % normal goat serum and 0.02% sodium azide in PBS (no detergent).

(E) Prepare the Zenon Reagents

For final anti-Ag A dilution of 1:400 and anti-Ag B dilution of 1:100 (Table 12.3), mix 2.5 μl of the mouse anti-Ag A with 20 μl of Fab 488 fluorophore and 10 μl of PBS followed by incubation for 5 min with agitation. Add 10 μl of blocking normal mouse serum and incubate with agitation for 5 min. The reaction mixture is brought to 500 μl. Next mix 10 μl of the mouse anti-Ag B with 20 μl of Fab 555 fluorophore and incubated for 5 min with agitation. Add 10 μl of blocking normal mouse serum and incubate with agitation for 5 min. The reaction mixture is brought to 500 μl. Mix the two 500 μl reagents and use immediately.

(F) Incubate with Labeled Antibody(ies)

Incubate cultures for 1 h with the labeled 1° antibodies mixed in the previous step.

(G) Rinse After Antibody Incubation

Use rinse buffer solution containing 10 mg/ml BSA and 5.0 % normal goat serum in PBS and rinse six times

(H) Fix with 4% Paraformaldehyde

A second fixation with 4% paraformaldehyde in PBS is needed for 30 min. This step will insure that the labeled Fab remains bound to the correct 1° antibody.

(I) Rinse after Antibody Incubation

Use rinse buffer solution containing 10 mg/ml BSA and 5.0 % normal goat serum in PBS and rinse six times

(J) Mount Coverslip

To remove the coverslips from the wells of the 24-well plate, use Dumont 3C forceps with tape on one side and a long syringe needle with bent at a right angle near the end. Use the bent syringe needle to lift the edge of the coverslip and to hold it off the substrate. Next open the forceps and grab the edge of the coverslip. Always hold the coverslip in 3C forceps so that the cells are on the side with the tape label you have previously added. Put a drop of mount medium on Parafilm, which is taped to the countertop. Place the coverslip tissue side down to float on the drop of mounting medium for 5 min. Remove the coverslip drain excess medium and place on a clear glass labeled microscope slide. Do not attempt to remove mounting medium of the noncell surface of the coverslip facing up until the medium has hardened in 24 h.

(K) Examine in Microscope

With wide field fluorescence microscope, examine the cultures for fluorescence with a 488 band pass filter set and with a 546 long pass filter set (Chapter 13, configure the software and hardware for the two fluorochromes using a band pass filter for examining the 488 nm fluorophore; explanation in Chapter 13, Microscopy and Images).

(L) Evaluate Results

Initially, look at an experimental to determine whether the 2° antibody labeling is as expected. Next, examine the 2° antibody control section to confirm no labeling. Finally, determine that there are no other labeling problems (e.g., background labeling) and that the level of labeling is acceptable. If there are problems with these conditions, then repeat the experiment with corrective measures (Chapter 14, Troubleshooting). Otherwise proceed to evaluate the results of the experiment and determine if the results meet your scientific expectations.

Chapter 13
Fluorescent Microscopy and Imaging

Keywords Immunohistochemistry · Antibody labeling · Fluorescence microscopy · Fluorescent immunocytochemistry · Fluorescent immunohistochemistry · Indirect immunocytochemistry · Immunostaining

Contents

Introduction   139
Filter Sets in Fluorescence Microscopy   140
Fluorescent Bleed-Through   142
Fluorescence Quench and Photobleach   145
Image Parameters – Contrast and Pixel Saturation   146
Ethics of Image Manipulation   148
    Do   149
    Do Not   149

Introduction

With the immunocytochemistry protocol completed, the tissue or cell preparation is now ready to be examined with a microscope. The type of microscope
and the location of the microscope were determined in Chapter 9 when the
Immunocytochemistry Experimental Design Chart and *Section 6, Microscopy,* were
completed.

The goal of this chapter is to understand the filter sets used for fluorescent microscope so that the best image can be collected. The upright microscope is the most
common choice for examining fixed tissue. The bright field microscope used to
examine labeling with HRP and its developed chromogens are familiar to most
scientists and will not be discussed.

R.W. Burry, *Immunocytochemistry*, DOI 10.1007/978-1-4419-1304-3_13,
© Springer Science+Business Media, LLC 2010

Filter Sets in Fluorescence Microscopy

The theory of fluorescence was presented in Chapter 6, Labels. To review, fluorescence is the property of a molecule that allows it to absorb one wavelength of light and emit a different higher wavelength. In the fluorescence microscope, the filter set is the important element, because it separates the excitation and the emission wavelengths.

The *fluorescence microscope* illuminates the sample from above. This form of illumination is called *epifluorescence* to distinguish it from *transmitted illumination* from below, which is the standard illumination in bright field microscopy. The key element of the fluorescence microscope is the *dichroic mirror*, which allows separation of excitation wavelengths from emission wavelengths (Fig. 13.1). The excitation light is reflected and travels to the sample. The emitted wavelengths pass through the objective lens and through the dichroic (Fig. 13.1). The dichroic, sometimes called the "beam splitter", allows the fluorescence light source to be above the sample because it reflects the excitation light to the sample. The parts of the *filter set* (Fig. 13.1) are the *excitation filter* (near the light source), which eliminates photons except those needed to excite the fluorophore, and the *emission filter* (above the

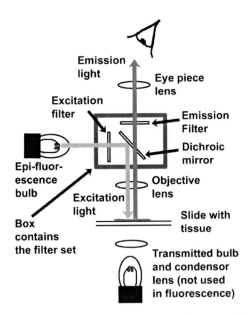

Fig. 13.1   Fluorescent microscope. Schematic drawing of a microscope with epifluorescence. The bulb and condenser at the *bottom* is used only for examining the slide with transmitted light. The fluorescent light path starts with the epifluorescence bulb (*left*), whose light (*light gray arrow*) passes through the excitation filter, being reflected down by the dichroic mirror, passes through the objective lens and onto the tissue section. The emitted light from the fluorophore (*dark gray arrow*) travels through the objective lens, the dichroic mirror, the emission filter, and the eye piece to the eye. The box is the filter set and contains the excitation filter, the dichroic mirror, and the emission filter.

dichroic), which allows only photons emitted from the fluorophore to be seen and the dichroic mirror.

The function of a filter set is more evident in the spectral graphs of the light at different wavelengths (Fig. 13.2). First, *the spectrum for visible light begins with blue at about 400 nm, green at 480 nm, orange at 580 nm, and red at 640 nm.* In Fig. 13.2a, a filter set for Alexa Fluor 488 has an excitation maximum at 495 nm (Fig. 13.2a, *black dashed line*) and emission maximum at 518 nm (Fig. 13.2b, *black dashed line*) are used. Even though the maxima for each wavelength is given as a single number, that number actually represents a bell-shaped curve with considerable excitation or emission occurring from 5 to 50 nm on either side of the maximum.

The excitation filter is located just before the dichroic (Fig. 13.1) and it allows passage of a narrow band of light from just under 450–500 nm (Fig. 13.2a, *solid gray line*). The graph of the excitation spectra of Alexa Fluor 488 extends beyond the

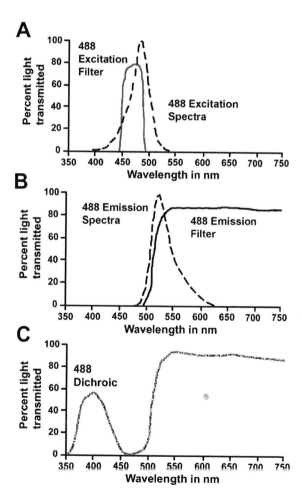

Fig. 13.2 Fluorescent filter set for a 488 fluorophore. The vertical axis indicates the amount of light transmitted at each wavelength. (**a**) The excitation filter is the solid *gray line* and the excitation spectra of the 488 fluorophore are the *dashed line*. The filter allows most of the excitation light to pass through. (**b**) The emission filter is the solid *black line* with the emission spectra of the 488 fluorophore are the *dashed line*. (**c**) The dichroic mirror showing transmission from 350 to 450 nm and from 500 nm and above. Between 450 and 500 nm, there is no transmission of light and the light is reflected.

capabilities of the filter, however, most of the excitation wavelengths are transmitted by the filter (Fig. 13.2a). The emission filter is located just after the dichroic (Fig. 13.1) and restricts light to only the wavelengths at or above 500 nm (Fig. 13.2b, *solid dark line*). The excitation wavelengths of Alexa Fluor 488 are transmitted through this filter (Fig. 13.2a). The dichroic must reflect light below 500 nm (Fig. 13.2c, *stippled line*) and transmit the light above 500 nm (Fig. 13.2c, *stippled line*). Looking at the graph of transmitted light for the dichroic (Fig. 13.2c), the *line* is near the *x*-axis from 450 to 500 nm, indicating no transmission of light. Above 500 nm, transmission is near 100% (Fig. 13.2c).

A complete view of the filter set shows the excitation filer, dichroic, and the emission filter spectra plotted together (Fig. 13.3a). This filter set is called *long pass filter set* because it will transmit all excitation photons above 550 nm. This long pass filter set allows maximum detection of the 488 emission.

For samples with multiple fluorophores, the fluorophores with lower excitations can be potentially detected by the filter sets for higher wavelength fluorophores, which means that long pass filter allows light with too high wavelengths to pass through. Instead, use a *band pass filter set* that allows only a narrow band of excitation to be observed (Fig. 13.3b). The range between 510 and 550 nm is the maximum emission for Alexa Fluor 488, but eliminates possible emissions from fluorophores with higher emissions that could be detected by the long pass filter set. While long pass filters are ideal for single fluorophore label in samples, almost all of the samples examined today have at least two and many times more fluorophores. Use only band pass filter sets for samples with multiple fluorophores. When bleed-through occurs, it is important to check the filter sets used to confirm they are band pass.

Additional fluorophores can be added to experiments and the filter sets selected accordingly. Typically, a green fluorophore (Alexa Fluor 488) and a red fluorophore (Alexa Fluor 555) are commonly used. For Alexa Fluor 555, with excitation maximum at 552 nm and an emission maximum at 568 nm, the filter set (Fig. 13.3c) has corresponding excitation and emission filters. Note, compare the filter set for Alexa Fluor 488 (Fig. 13.3b) with that for Alexa Fluor 555 (Fig. 13.3c), and confirm that the filters and dichroics allow different wavelength to be transmitted. For most samples, these two filter sets will allow collection of images without bleed-through.

Selecting fluorophores for experiments with multiple labels requires that there is enough separation between the excitation and emission wavelengths to prevent bleed-through. In the example above, the separation between the excitation maxima is 80 nm. If the excitation maxima of fluorophores are separated by 80–100 nm, then four fluorophores can be safely used in the visible range. An ideal spread for fluorophores is 350, 488, 546, and 744. Examples of fluorophores are given in Table 13.1.

Fluorescent Bleed-Through

Bleed-through occurs when examining two fluorophores and the fluorophore with lower excitation and emission spectra will appear as an emission of a higher

Fig. 13.3 Fluorescent filter sets long pass and band pass. (**a**) The combined components of the 488 excitation filter (*gray line*), the emission filter (*black line*), and the dichroic (*stippled line*) as seen in Fig. 13.2. This is a long pass filter set because it captures all of the emission light at higher wavelengths. (**b**) When using multiple fluorophores, a band pass filter is needed to collect emission light over a narrow range of wavelengths. The emission filter transmits light only between 500 and 550 nm. (**c**) A band pass filter for the 555 fluorophore has different filters and dichroics that detect light only between 575 and 630 nm.

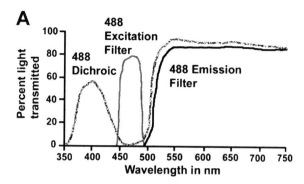

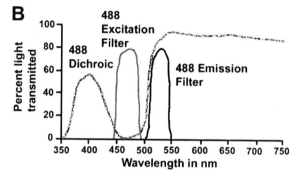

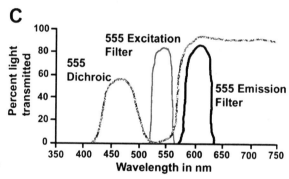

Table 13.1 Combinations of multiple fluorophores

350 nm (blue)	DAPI, AMCA, Alexa 350, Fast Blue, Fluoro-Gold, Hoechst,
488 nm (green)	FITC, Cy2, Alexa 488, GFP, YFP
546 nm (red)	Rhodamine, Cy3, Alexa 546, DiI, DsRed
647 nm (high red)	Cy5, Alexa 647, DRAQ5

wavelength fluorophore. This situation only occurs when the intensity of the lower wavelength fluorophore is very high and the intensity of a higher wavelength fluorophore is very low. Remember that the emission spectrum of a fluorophore has broad tails that are not important if the two fluorophores label with near-equal levels. When the lower wavelength fluorophore has very high intensity levels its higher wavelength tail can be detected. In Fig. 13.4, the maximum intensity of the 488 emission spectrum is very high and the maximum intensity of the 555 emission spectrum is low. This allows the tail of the 488 emission spectrum (Fig. 13.4 dark gray, Bleed-through) to be half the intensity of the 555 emission intensity. With the 555 emission filter (Fig. 13.4, gray area under 555 emission filter curve), about a third of the total photons detected will come from the 488 emission tail that extends into the filter. When the weak 555 signal is amplified, the amplification will also increase the 488 emission signal.

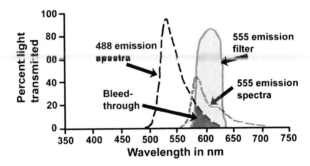

Fig. 13.4 Fluorescent bleed-through. A problem in fluorescent microscopy occurs when the filters cannot separate emission wavelengths, for example, when an emission from 488 fluorophore is detected in the 555 channel and in the 555 emission. The emission spectrum for the 488 fluorophore (*black dashed line*) is shown with a tail that extends into the 555 emission filter (*gray solid line*). If the intensity of the 555 fluorophore emission is low as seen by the maximum height at 40% and the 488 fluorophore has high-intensity emission, then the amount of overlap in emission can be significant (*dark gray area*). To compound the problem, if the detection system gain is increased in the *red* 555 emission, then the 488 fluorophore intensity is about 50% of the light as from the 555 fluorophore intensity.

To demonstrate bleed-through, cells are labeled with phalloidin 488 fluorophore, which binds to actin microfilaments, and a mitochondrial dye 555 fluorophore. When viewed individually without bleed-through, the green actin filaments are seen as linear structures (Fig. 13.5a). The red mitochondrial label is punctate seen mainly around the unlabeled nucleus (Fig. 13.5b). The merged micrograph shows the distinct green and red colors with no colocalization or yellow (Fig. 13.5c). When this cell is viewed with both fluorophores collected together, the bleed-through is seen. The green actin filaments are seen as before (Fig. 13.5d) with a bundle of filaments seen at the *arrow*. Now, the red channel shows in addition to the expected mitochondria actin filaments (Fig. 13.5e, *arrow*). In the merged micrograph, the actin filaments are now yellow, indicating colocalization of both the green and red channels.

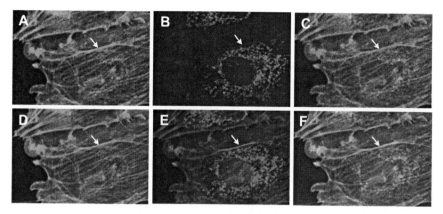

Fig. 13.5 Fluorescent bleed-through micrographs. Micrographs (a)–(c) show fluorescent cells with no bleed-through and micrographs (d)–(f) show the same field with bleed-through. (a) Phalloidin with 488 fluorophore labeling actin. The *arrow* shows one spot of high-intensity label with the 488 fluorophore that is not labeled with the 456 fluorophore. (b) The dye, Mitotracker *Red* (Molecular Probes/Invitrogen), labels only mitochondria. (c) The merged image shows no *yellow* colabeled structures and only *green* labeled actin with *red* labeled mitochondria. (d) In a new set of micrographs of the same cells, using Phalloidin with 488 fluorophore labeling actin. (e) The dye Mitotracker *Red* labels mitochondria and in this case, the *red* label in addition to the mitochondria, corresponds to the actin filaments as seen at the *arrow*. (f) The merged image shows the double-labeled structures in *yellow* with no label specific to the *green* channel. In this case all of the *green* label would be incorrectly thought to colabel with the *red* label. Bleed-through is always from a lower wavelength fluorophore to a higher wavelength fluorophore.

With wide field fluorescence, it is not possible to separate these two fluorophores. However, with the confocal microscope, it is possible to excite each fluorophore individually with a separate laser and collect the emissions as sequential or multi-track images. To fix bleed-through, reverse the 1° antibodies so that the highest intensity signal is in the highest wavelength fluorophore and the lowest intensity signal is in the lowest wavelength fluorophore. The best way to identify bleed-through is to examine samples labeled with only one fluorophore and determine that the distribution of the label is the same as that in the multiple fluorophore samples.

Fluorescence Quench and Photobleach

Fluorophores examined in a fluorescence microscope fade with increased viewing time due to one of the two causes, quenching or photobleaching. *Quenching* is the reduced output of fluorophore because its emission energy is absorbed by adjacent molecules, such as oxidizing agents, salts, heavy metals, or other fluorochromes. In immunocytochemistry, quenching is not a problem because mounting medium does not contain quenching molecules.

Photobleaching is the reduced output of a fluorophore due to irreversible damage to the molecule in the presence of molecular oxygen. This is a permanent loss of the fluorophore molecule and is a significant problem for immunocytochemistry. There are three ways to address photobleaching.

First, reduce photobleaching by reducing exposure of fluorophores to excitation light. While this sounds obvious, it requires thought when examining fluorescent-labeled samples. Confine the overview examination to a minimum amount of time and shut off the light path to the sample when not looking at the sample. Another good method is to use lower magnification lenses or no zoom when finding a sample and adjusting the image collection device. Then increase magnification only for collecting the final image.

Second is the historical approach using antifade agents, also called antibleach agents. *Antifade agents are reducing compounds that lower the amount of free oxygen in the mounting medium and reduce the chance of photobleaching.* Initially, compounds like para-phenylenediamine were used, but within a few weeks, these compounds become fluorescent themselves thus increasing the background in the sample. Some of the newer antifade compounds do not become fluorescent over time.

The third way to reduce photobleaching is to use different fluorophores dyes that resist photobleaching. FITC and rhodamine are first-generation fluorophores and fade very rapidly. Their use today for immunocytochemistry is not recommended. The second generation of fluorescent compounds, the Cy dyes, are better at resisting photobleaching. The third generation of fluorescent compounds, are one of the best at resisting photobleaching. Third-generation dyes can be examined for minutes at reasonable levels of illumination with minimal photobleaching. Nanoparticles (Qdots) do not photobleach and in theory should be the fluorophore of choice. However, their large size and difficulty of penetration into cells show that they are not recommended for immunocytochemistry at this time.

Image Parameters – Contrast and Pixel Saturation

With the field selected, the next important step is setting the digital camera or the confocal detector. Incorrect settings result in poor-quality image, even though the sample is well prepared. The most important settings for image collection are controls for image brightness and contrast (sometimes called black levels).

With a digital camera or a confocal microscope (also digital), images are collected in arrays of pixels, which represent the image in the height and width. Today, devices collect standard images of 512×512 with the details of the entire image being divided into 262,144 individual pixels. There are higher levels of resolution, like 1024×1024, but these will take longer to collect an image. Each pixel has a capacity or depth to indicate the intensity of the signal for that pixel. This is called pixel gray levels and they range from black to white. An 8-bit pixel depth has 256 gray levels of intensity, a 12-bit has 4,096 gray levels, and a 16-bit has 65,536 gray levels.

A standard image (Fig. 13.6a only partially shown here) has 512×512 pixels and a 12-bit pixel depth. The white horizontal *line* shows the location where

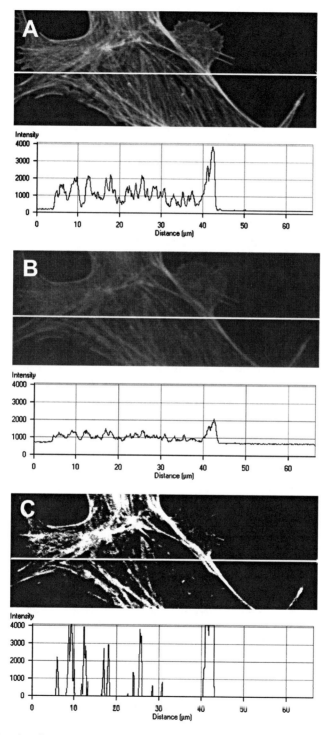

Fig. 13.6 (continued)

the signal intensity was measured (see graph immediately below each image). This graph shows the intensity of the emission at each pixel on the vertical axis and the position across the image in the horizontal axis. The 4096 intensity levels indicate that this image is a 12-bit image. The contrast is evaluated by looking at highest pixel intensity and the levels of the background intensity. This is an example of a good image with balanced contrast, which provides a maximum amount of information because the most intense pixels are near 4096 and the black/blank pixels are above zero.

The examples in the other images are poor quality images that exclude or reduce information from the image. A low contrast image (Fig. 13.6b) looks washed out and the background is not black. The graph of the pixel intensities for this image shows that the highest pixels are under 2000, which is not taking advantage of the full 4096 range of the detector. Also the background intensity is just under a 1000, which does not leave the background black but gives the background some visible intensity. The low-contrast image not only does not look good, but does not take advantage of the intensity range of the detection device and therefore does not show all the information that is contained in the image.

The high-contrast image (Fig. 13.6c) clearly shows some of the structures in the image, but it does not show the fine detail of the image seen in the best exposure (Fig. 13.6a). Looking at the graph for this image, many pixels have intensity values that are too high, as indicated by the flat tops at the top of the graph. These high pixels are "saturated" because they show complete white and do not give a gray range. At the bottom of the graph, most pixels have no value indicating they are empty. This image is not showing the complete range of gray necessary for a good quality. There should be intensity in all pixels that indicate a potential structure. Some scientists submit images that are set at this level of high contrast to make the results more dramatic, almost black and white. Most journals will not accept high-contrast images because they are considered distorted.

Ethics of Image Manipulation

Once images have been captured and viewed, selections are prepared for publication. With easy-to-use software packages, micrographs can be enhanced or edited.

Fig. 13.6 (continued) Fluorescent micrograph contrast and saturation. In taking a micrograph, the adjustments are important for collecting as much information as possible from the cells. (a) Correctly exposed micrograph of Phalloidin with 488 fluorophore labeling actin in cells. The *white line* indicates where the intensity graph below was made. Note the most intense levels are under the maximum intensity of 4096 and the background is slightly above the zero level. (b) An underexposed micrograph that appears washed out with background that is not *black*. The intensity graph shows the most intense label is not near the maximum intensity and the background has a level substantially above zero. (c) This micrograph has too much contrast. The image is basically *black* or *white* with no *gray shades*. The intensity graph shows that much of the label is at the maximum intensity with flat peaks at the *top* of the graph, and the *lower* view shows no intensity above zero.

Years ago, when only film and photographic paper were used to capture images, there were very few things a scientist could do to change a micrograph. Today, manipulating images poses a potential ethical problem because it is possible to grossly alter the images so that they are different from the original. Digital images can be processed or edited. However, only certain alterations to images are ethical in preparation for publication. There have been a large number of excellent publications that discuss manipulation (Rossner and Yamada, 2004; North, 2006; Pearson, 2007). Presented below is a selected summary of the most important points.

The basic point is that all images must accurately represent the data originally collected. Misrepresentation of data is deception and violates the concept of scientific honesty that is expected in all scientific work. If images are not perfect but accurately represent what is found, then they should be used to illustrate the point.

Do

- Keep an archive of the original, unchanged image as collected. Some journals will request original images before a manuscript is published.
- Adjust brightness or intensity identically for all images, whether they are experimental and control.
- Report all image manipulations in the Materials and Methods or figure legends of any submitted manuscript.
- Adjust all images to be compared in a single figure the same way and to the same extent.
- Use pseudocolor to enhance differences in gray scale
- Add a scale bar (micron bar) to each image or for each group of images at the same final magnification.

Do Not

- Cut, add, duplicate, remove, or move part of an image to or from the same image or another image. The image must always be treated as a whole.
- Treat control images differently from experimental images.
- Change the resolution of the image (size of a pixel) so that it apparently adds resolution to the micrograph.
- Perform nonlinear adjustments (i.e., gamma) to images that will emphasize one aspect and/or remove another aspect of an image.
- Magnify differently individual images that are part of a group of images to be compared

Chapter 14
Troubleshooting

Keywords Immunohistochemistry · Antibody labeling · Fluorescence microscopy · Fluorescent immunocytochemistry · Fluorescent immunohisto-chemistry · Indirect immunocytochemistry · Immunostaining

Contents

Introduction 151
Procedural Errors 152
Method of Troubleshooting 152
   Case No. 1 153
   Case No. 2 156
   Case No. 3 158
   Case No. 4 164
   Case No. 5 167
Troubleshooting Unique to Multiple Primary Antibodies 173
Bad Antibodies 173
   Bad 1° Antibodies 173
   Bad 2° Antibodies 174

Introduction

Most of the immunocytochemistry problems seen in a shared use microscope facility are problems with the immunocytochemical procedure. In immunocytochemistry problems will frequently occur and need to be solved by troubleshooting. Most of the method problems come from three sources: (1) modified immunocytochemistry protocol that was not totally re-evaluated after being modified; (2) protocol designed without sufficient thought, generally by a novice; and (3) procedure errors where the protocol was not followed.

Sources of most common problems (not in any order):

- Fixation or procedural errors
- Failure to find optimal antibody concentrations (more is not always better)

R.W. Burry, *Immunocytochemistry*, DOI 10.1007/978-1-4419-1304-3_14,
© Springer Science+Business Media, LLC 2010

- Species of 2° antibody do not match species of 1° antibody
- Cross-reactivity of antibodies in multiple labeling
- Blocking inappropriate (too little or the blocking binds to antibodies)
- Use of inappropriate filter sets and/or laser lines in microscopy
- Attempts to overcome problems with image collection methods (increased amplification of intensity)

Procedural Errors

When faced with an immunocytochemistry problem, check the simple things first. These are the steps in the procedure and confirm they were done as designed. In many cases, problems resulted not following the procedure. For example, if a person is interrupted while performing a procedure when the procedure was resumed, perhaps, steps were skipped or repeated. When in a hurry, it is easy to grab a tube or vial and not read the label. Sometimes in labs, reagents can be contaminated, and this is very hard to detect.

- Confirm that none of the reagents was omitted (1° antibody, 2° antibody, and blocking, detergent).
- Confirm that reagents were added in the correct order and for sufficient incubation times.
- Reread labels to confirm that the correct antibodies and reagents were used. This is especially important when using multiple 1° and 2° antibody combinations.
- Check antibody titrations and dilutions to insure they were done properly.
- Confirm that the correct filter set (and laser lines) was used to collect the images.

Method of Troubleshooting

Sherlock Holmes was a master at troubleshooting. From reading the accounts of his work, it becomes clear that there are three steps in solving a crime or troubleshooting problems with immunocytochemistry:

(1) First, *define the problem* of the mysterious crime, whether it is a murder, theft, or a missing person.
(2) Second, collect the available information by listening to the stories of the events. Holmes always arrived at a series of explanations that could have led to the crime by use of *deduction or hypothesis generation*.
(3) Third, *investigate* and it is this exciting detective work that takes up most of the enthralling part of the story and involves interviews, disguises, and chemistry experiments, followed by an "ah-ha" moment. To solve the problem requires collection of additional data to select the correct explanation.

The method of identifying and solving immunocytochemistry problems in images is an exercise in problem solving or the Sherlock Holmes method. The steps are

First, *define the problem* such as the label is missing or inappropriate.

Second, *generate hypotheses* that have as many of the potential causes of the problem as possible.

Third, *investigate* each potential cause. First use controls and then conduct additional experiments as needed.

The following is a series of examples from experiments performed to mimic real experiments. Initially, the micrographs show the problem and allow you to *define the problem*. To *generate hypotheses* will involve understanding the potential causes that could give rise to the problem. This is not an exercise in thinking of as many possibilities as possible, but rather only thinking of those possibilities that could have given rise to the problem. If you have little experience with immunocytochemistry, you will have many individual hypothesis that will need to be eliminated. *Investigate* is the step where additional experiments will be needed to solve the problem. The examples here may seem simple when presented, but when you are faced with a problem, no solution is simple or obvious. For the examples that follow, the use of the Experimental Design Chart would have eliminated the causes of problems. In fact, to solve these problems you might try to fill out the Experimental Design Chart to see if you find the problem.

Case No. 1

Experiments were done to determine the distributions of ribosome proteins in mouse spinal cord following various treatments. In the initial experiment rabbit anti-ribosomal protein antibody was used at 1:1000 and goat anti-rabbit 488 fluorophore at 1:1000. In the wide field fluorescent micrographs, the specific label was low (Fig. 14.1a, *arrows*) and the background was very high. The no 1° antibody control has a similar level of background (Fig. 14.1b). The person who did the experiments said that the background was high because the specific labeling could not be easily seen and tried to work on the images with a computer imaging program.

Define the Problem

The micrographs clearly show high background (Fig. 14.1b), but they also show that the specific labeling was not very intense (Fig. 14.1a, *arrows*). While the person who did this experiment focused on the high background, the major problem is the low level of specific labeling. The procedure was as follows

Tissue was from a mouse perfused with 4% paraformaldehyde made the same day.

Infiltration was done with 30% sucrose in PBS.

Cryostat sections were mounted on microscope slides.

Permeabilization was done with 0.3% Triton X-1000 in PBS containing 5% normal goat serum, 10 mg/ml of BSA.

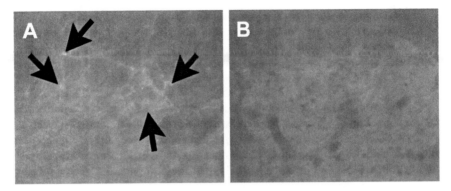

Fig. 14.1 Case No. 1 problem. Tissue is from the mouse spinal cord incubated with rabbit anti-ribosomal protein antibody at 1:1000 and goat anti-rabbit 488 at 1:1000, examined in a wide field fluorescence microscope. (**a**) This micrograph shows extremely light specific label (*arrows*) and a very high background. (**b**) A no 1° antibody control with high background and no specific labeling are shown

Rabbit anti-ribosomal protein 1:1000 in PBS contained 5% normal goat serum, 10 mg/ml of BSA.

Rinses were with PBS containing 5% normal goat serum, 10 mg/ml of BSA.

Goat anti-rabbit 488 fluorophore 1:1000 in PBS containing 5% normal goat serum, 10 mg/ml of BSA.

Rinses were with PBS contained 5% normal goat serum, 10 mg/ml of BSA.

Imaging was done on a wide field fluorescence microscope with a 40× air N.A. 0.8 objective and a CCD camera with 4–5 s exposures.

Generate Hypotheses

Hypothesis generation will include things that could give low levels of specific labeling or high levels of background label. The following is a potential list of problems:

(1) Procedural errors
(2) 1° antibody not specific for antigen
(3) 2° antibody not correct for 1° antibody species
(4) Insufficient or wrong blocking agents
(5) Insufficient rinses
(6) Fluorescent filter set did not match the fluorophore
(7) Photobleaching of specific label
(8) Dilution of 1° and 2° antibody incorrect
(9) Incorrect collection of micrographs with CCD camera

Investigate

The first thing to check is that all the controls were performed. The results show that a no 1° antibody control gave a high level of fluorescence similar to the sample incubated with the 1° antibody. Also the blocking agents were similar to those used successfully in other experiments.

(1) Procedural errors – Checking of solutions and dilutions showed no procedural errors.
(2) 1° antibody not specific for antigen – Literature from the vendor showed an immunoblot.
(3) 2° antibody not correct for 1° antibody species – The rabbit anti-ribosomal protein 1° antibody was used with a goat anti-rabbit IgG labeled with 488 fluorophore 2° antibody.
(4) Insufficient or wrong blocking agents – Blocking with 5% normal goat serum the same species as the 2° antibody, 10 mg/ml of BSA were sufficient for other rabbit 1° antibodies.
(5) Insufficient rinses – Seven rinses were used between the antibodies and before mounting.
(6) Fluorescent filter set did not match the fluorophore – A check of the microscope showed the filter set was correct.
(7) Photobleaching of specific label – Brief exposure of the excitation light on the sample was used in taking the micrographs
(8) Dilution of 1° and 2° antibody incorrect – The dilutions of the 1° and 2° antibodies had not been tested rather dilutions were selected from those recommended. To investigate the antibody dilutions, a Dilution Matrix was done, and the results are shown in Fig. 14.2. These results show that at the dilutions

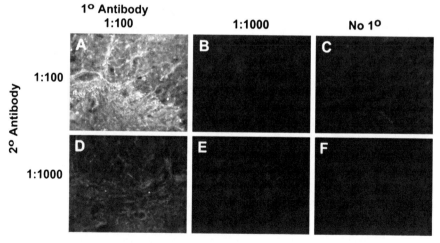

Fig. 14.2 Case No. 1 solution. A dilution matrix experiment with the rabbit anti-ribosomal protein antibody. (**a**) With 1° antibody 1:100 and 2° antibody at 1:100, specific labeling was seen with high background. (**b**) With 1° antibody 1:1000 and 2° antibody at 1:100, no specific labeling was found. (**c**) With no 1° antibody and 2° antibody at 1:100, no specific labeling was found. (**d**) With 1° antibody 1:100 and 2° antibody at 1:1000, slight specific labeling was seen. (**e**) With 1° antibody 1:1000 and 2° antibody at 1:1000, no specific labeling was seen. (**f**) With no 1° antibody and 2° antibody at 1:1000, no specific labeling was seen

used in the experiment, 1° antibody 1:1000 and 2° antibody 1:1000, were inappropriate and weak fluorescence was detected (Fig. 14.2e). With 1° antibody at 1:100 and 2° antibody at 1:100, the best labeling was seen (Fig. 14.2a), but these with dilutions so low that background was seen (Fig. 14.2c). It is clear that with this 1° antibody dilution and the background level a different 1° antibody should be found for the experiment. In the results from the Dilution Matrix, the labeling in the no 1° antibody control showed no background, unlike the results originally obtained with the experiment.

(9) Collection of micrographs with CCD camera – The images in the original experiment were collected at long exposure times, followed by computer contrast enhanced. The images in the Dilution Matrix were collected at shorter exposure times so that the background was always lower.

There were two problems in Case No.1. First, the 1° and 2° antibody dilutions were too high so that it was almost impossible to find any signal in the slides. The second problem was that in an attempt to correct for the low antibody dilutions, the micrographs were manipulated after they were collected to enhance the contrast. The person who performed this experiment did enhance both the experimental and control images the same way so that they accurately represented comparable images. The correct way to perform the experiment was to use 1° and 2° antibody dilutions at 1:100 and to collect images with contract and brightness set to show a black background.

Case No. 2

Immunocytochemistry was performed with a 1° antibody for a synaptic vesicle protein p38 that gives fine punctate labeling in the central gray matter of the spinal cord and a small amount of label in small strands in the white matter. As seen with a confocal microscope (Fig. 14.3a), the specific labeling looks as expected with the gray matter to the lower right and the unlabeled white matter to the upper left. The *asterisk* (Fig. 14.3a) indicates a high background in an area of no labeling of the white matter. The high background is confirmed with a no-primary antibody control (Fig. 14.3c), which also shows high levels of nonspecific background labeling. There is labeling on the microscope slide in some areas that contained no tissue (Fig. 14.3b). The edge of the tissue is indicated by the *dashed white line*, with the tissue to the lower right and bare microscope slide to the upper left. The *arrows* (Fig. 14.3b) show unexpected labeling on the microscope slide. This pattern of labeling was also seen on the tissue to the lower right, which should not have labeling.

Define the Problem

Single antibody experiment shows high background with labeling on some areas of the microscope slide. The procedure was as follows

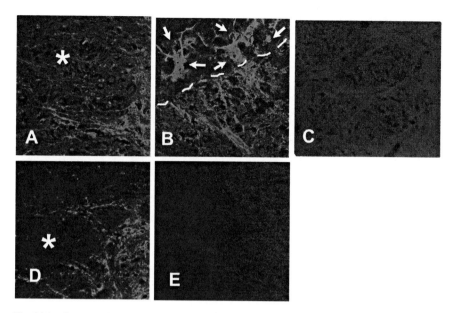

Fig. 14.3 Case No. 2. Mouse spinal cord incubated with mouse anti-p38 antibody used at 1:1000 and goat anti-mouse 546 at 1:1000 is examined in a confocal microscope. (**a**) Specific labeling was high and high background was seen in some, but not all areas that should not be labeled (star). (**b**) Also, in one area at the edge of one section (*white dashed line*), where there should be no labeling on the tissue, showed labeling on the tissue (to the *lower right* of the *white dashed line*) and on the glass of the microscope slide (*arrows*). (**c**) The no 1° antibody control showed high background. The problem in these sections was that the 1° antibody dried on to the sections. (**d**) In a repeat experiment, where the sections did not dry, specific labeling was high and there was no background. (**e**) The no 1° antibody control showed no background when the experiment was repeated

Spinal cord tissue was from a mouse perfused with 4% paraformaldehyde made the same day.

Infiltration was done with 30% sucrose in PBS.

Cryostat sections were mounted on microscope slides.

Permeabilization was done with 0.3% Triton X-1000 in PBS containing 5% normal goat serum, 10 mg/ml of BSA.

Mouse anti-p38 dilution was 1:1000.

Rinses with PBS contained 5% normal goat serum, 10 mg/ml of BSA.

Goat anti-mouse IgG 488 fluorophore was 1:1000.

Rinses with PBS contained 5% normal goat serum, 10 mg/ml of BSA.

Imaging was done with a 63× oil N.A. 1.43 objective on a confocal microscope.

Generate Hypotheses

The appearance of this labeling in the experimental and controls suggests a number of potential problems. The following is a potential list of problems.

(1) Procedural errors
(2) Insufficient or wrong blocking agents
(3) Insufficient rinses
(4) Dilution of 1° and 2° antibody
(5) 2° antibody is bad
(6) Solutions dried on sections

Investigate

(1) Procedural errors – All the solutions were correct.
(2) Insufficient or wrong blocking agents – Examination of the procedure showed that standard block agents were used and these had worked previously.
(3) Insufficient rinses – The rinses were done with seven after each antibody.
(4) Dilution of 1° and 2° antibody – A Dilution Matrix was performed previously and so the dilutions of the 1° and 2° antibodies were correct.
(5) 2° antibody is bad – The 2° antibody was tested with a different mouse 1° antibody and worked well.
(6) Solutions dried on sections – A check for solutions drying on the slide should be a part of the examining the procedural errors. In this case, a discussion with the person revealed that for the overnight 1° antibody incubation he forgot to put water in the chamber holding the slides and the slides did dry out with 1° antibody drying on to the tissue. The person said that extra rinses were used and they thought the antibody would be washed off.

The problem in case No. 2 was that the 1° antibody incubation solution dried on the section. The micrograph (Fig. 14.3b, *arrows*) showed the dried antibody on the microscope slide next to the tissue. The 2° antibody, goat anti-mouse 488 fluorophore, then bound to the attached 1° antibodies. The dried antibody is not possible to rinse off slides of sections, and when it occurs, the slides should be discarded. When this experiment was repeated and when antibody did not dry out, the labeling was as expected (Fig. 14.3d) with low background (Fig. 14.3d, *asterisk*), that matched low labeling seen in the no 1° antibody control (Fig. 14.3e)

Case No. 3

To examine the distribution in the gray matter of the spinal cord of a glial cell type called oligodendrocyte, immunocytochemistry was performed. The first 1° antibody is against Oligo, a protein unique to oligodendrocytes. The second 1° antibody was against synaptic vesicle protein, p38. The first experiment confirmed the antibody dilutions for each 1° antibody singly and showed the demonstrated expected labeling for synaptic vesicle protein (Fig. 14.4a) and the oligodendrocytes (Fig. 14.4b, *arrows*).

In a double-label experiment, the results were not as expected. The p38 labeling was punctate as expected (Fig. 14.4c), but the Oligo labeling showed not only the lightly labeled cells (Fig. 14.4d, *arrows*), but also punctate labeling than in the merged image (Fig. 14.4e) co-localized with the p38, as seen by the yellow color. This result indicated that label for the p38 was also found with the Oligo protein labeling. The no 1° antibody controls showed the mouse anti-p38 label seen in both the expected green channel (Fig. 14.4f) and unexpectedly in the red channel (Fig. 14.4 g). However, the no 1° antibody control with goat anti-Oligo incubation, no labeling in the green channel (Fig. 14.4), but normal oligodendrocyte labeling was seen in the red channel (Fig. 14.4j).

Define the Problem

The mouse anti-p38 labeling is found in both the expected green channel and in the unexpected red channel. The procedure was a follows

> Tissue was from a mouse perfused with 4% paraformaldehyde made the same day.
> Infiltration was done with 30% sucrose in PBS, sections were cut on a cryostat.
> Permeabilization was done with 0.3% Triton X-100 in PBS containing 5% donkey serum, 10 mg/ml of BSA.
> Mouse anti-p38 1:1000 was used.
> Rinses with PBS contained 5% normal donkey serum, 10 mg/ml of BSA.
> Goat anti-mouse with 488 fluorophore was used.
> Rinses with PBS contained 5% normal donkey serum, 10 mg/ml of BSA.
> Goat anti-Oligo 1:200 was used.
> Rinses with PBS contained 5% normal donkey serum, 10 mg/ml of BSA.
> Donkey anti-goat IgG labeled with 546 fluorophore 1:1000 was used.
> Rinses with PBS contained 5% normal donkey serum, 10 mg/ml of BSA.
> Imaging was done with a 40× oil N.A. 1.2 objective on confocal microscope.

Generate Hypotheses

This experiment was performed differently from the methods used previously. The sets of 1° and 2° antibodies were processed in series with the first 1° antibody, mouse anti- p38, and the first 2° antibody, goat anti-mouse 488 fluorophore, completed before the series for the Oligo.

The appearance of this labeling in the experimental and controls suggests only a few potential problems. Where there is a problem with cross-reaction between the sets of antibodies, the following is a potential list of problems:

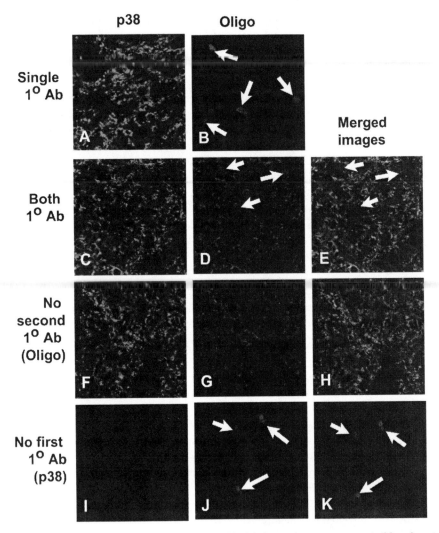

Fig. 14.4 Case No. 3 problem. In this serial double-label experiment, mouse anti-p38 and goat anti-mouse 488 fluorophore labeling was completed first and the goat anti-Oligo and donkey anti-goat 543 fluorophore followed. (**a**) In results from the dilution matrix, the mouse anti-p38 showed punctate labeling in the spinal cord. (**b**) Also from the dilution matrix, the goat anti-Oligo showed labeling for cells (*arrows*) in the spinal cord. (**c**) The double-label experiment showed mouse anti-p38 punctate labeling as expected. (**d**) In the same field, the goat anti-Oligo labeling showed low-intensity label over cell bodies (*arrows*) and unexpectedly particulate labeling. (**e**) Merged micrograph of (**c**) and (**d**) showed that all the p38 labeling was yellow; the cell bodies (*arrows*) were red. (**f**) Control with mouse anti-p38 antibody and no goat anti-Oligo antibody showed the expected punctate labeling in green. (**g**) In the same field, the missing goat anti-Oligo antibody had unexpected punctate red labeling. (**h**) Merged micrograph of f and g showed that all the p38 labeling was yellow. (**i**) Control with no mouse anti-p38 and only goat anti-Oligo antibody showed no labeling for p38. (**j**) In the same field, the goat anti-Oligo labeling showed only the expected cell body labeling (*arrows*). (**k**) Merged micrograph of i and j showed only the labeling for the cells.

(1) Procedural errors
(2) Incorrect blocking
(3) First 1° antibody was not specific
(4) First 2° was not specific
(5) Second 2° was not specific
(6) Species of the second 2° antibody was incorrect.

Investigate

The individual hypotheses were evaluated.

(1) Procedural errors – Checks of solutions and dilutions showed no errors.
(2) Incorrect blocking – The 2° antibodies were made in donkey and goat with the blocking serum made in donkey.
(3) First 1° antibody is not specific – Previous use of this antibody has shown it is specific.
(4) First 2° was not specific – Tests of this 2° antibody with other 1° antibodies show it is specific.
(5) Second 2° was not specific – Tests of this 2° antibody with other 1° antibodies show it is specific.
(6) Species of the second 2° antibody was incorrect – Examination of the species of the 1° antibodies indicates that the p38 was made in mouse and the Oligo was made in goat. The 2° antibodies for p38 was goat anti-mouse and the Oligo was donkey anti-goat. To help understand this problem, the binding of the mouse anti-p38 antibody to the p38 antigen was followed by incubation with goat anti-mouse 488 fluorophore (Fig. 14.5a). The problem is that both the second 1° antibody (goat anti-Oligo) and the first 2° antibody (goat anti-mouse) were made in goat. The second 2° antibody was a donkey anti-goat 543 fluorophore and bound to both of these antibodies (Fig. 14.5c). As a result, both the 1° antibodies were labeled with the donkey anti-goat fluorophore 546.

To correct this problem the species of the first 2° antibody was changed. In multiple 1° antibody experiment when using a 1° antibody made in goat (e.g., Oligo), it is not possible to use any 2° antibody made in goat. The rule is that the species of all the 2° antibodies cannot be the same as the species of any of the 1° antibodies.

The correct experiment was performed where the first 2° antibody was now a donkey anti-mouse 488 fluorophore. The results show the expected labeling for both antibodies together (Fig. 14.6a, b, c) the no 1° antibody controls also show labeling only for the correct protein (Fig. 14.6d, e, f, g, h, i).

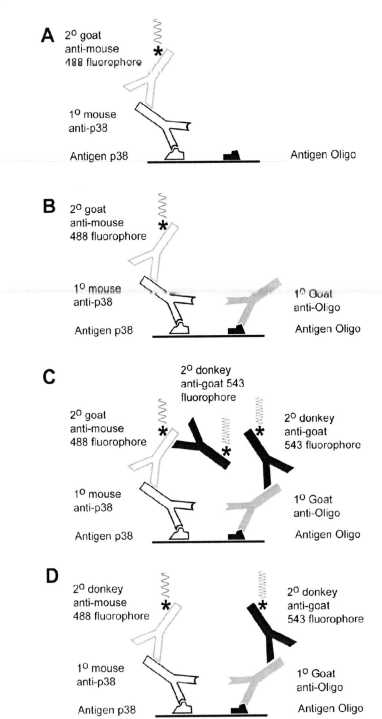

A   2° goat
anti-mouse
488 fluorophore

1° mouse
anti-p38

Antigen p38   Antigen Oligo

B   2° goat
anti-mouse
488 fluorophore

1° mouse
anti-p38   1° Goat
                                                       anti-Oligo

Antigen p38   Antigen Oligo

C   2° donkey
                           anti-goat 543
                           fluorophore

2° goat   2° donkey
anti-mouse   anti-goat
488 fluorophore   543 fluorophore

1° mouse   1° Goat
anti-p38   anti-Oligo

Antigen p38   Antigen Oligo

D   2° donkey   2° donkey
anti-mouse   anti-goat
488 fluorophore   543 fluorophore

1° mouse   1° Goat
anti-p38   anti-Oligo

Antigen p38   Antigen Oligo

Fig. 14.5 (continued)

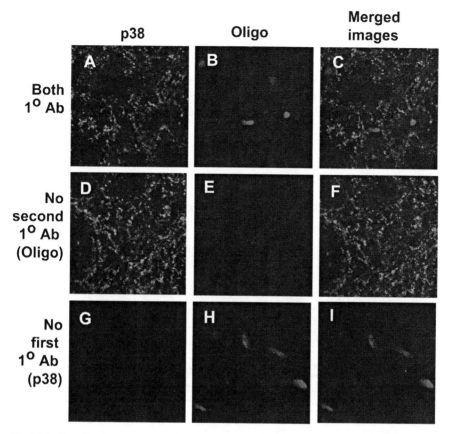

Fig. 14.6 Case No. 3 correct experiment. Changing the first 2° antibody from goat anti-mouse to donkey anti-mouse allowed the experiment to give expected labeling. (**a**) Mouse anti-p38 with donkey anti-mouse 488 fluorophore showed punctate labeling as expected. (**b**) Goat anti-Oligo and donkey anti-goat 546 fluorophore showed expected cell labeling. (**c**) Merged micrograph of (**a**) and (**b**) showed totally separate labeling as expected. (**d**) Control with mouse anti-p38 and no goat anti-Oligo antibody showed only the expected punctate labeling. (**e**) In the same field goat anti-Oligo antibody that was omitted and no labeling was seen. (**f**) Merged micrographs for (**d**) and (**e**). (**g**) Control with no mouse anti-p38 and only goat anti-Oligo antibody showed no labeling for p38. (**h**) In the same field, the goat anti-Oligo labeling showed only the expected cell body labeling. (**i**) Merged micrographs of (**g**) and (**h**) showed only the labeling for the cells

Fig. 14.5 Case No. 3 explanation. Incubations in the multiple-antibody experiment were done with the first set of 1° and 2° antibody completed before the second set. (**a**) Incubation of the mouse anti-p38 binding to the p38 antigen was followed by incubation with goat anti-mouse 488 fluorophore. (**b**) In the incubation with goat anti-Oligo antibody it bound to the Oligo antigen. (**c**) The incubation with the donkey anti-goat 543 fluorophore antibody bound to the goat anti-Oligo 1° antibody and the goat anti-mouse 488 fluorophore 2° antibody. (**d**) The correct experiment was done with a different first 2° antibody made in donkey, not goat

Case No. 4

To investigate neuronal synapses in the white matter of mouse spinal cord, immuno-cytochemistry with two 1° antibodies was performed. The mouse anti-p38 antibody labels synaptic vesicle protein, with a secondary labeled with a 543 fluorophore were previously used in the lab. The rabbit anti-neurofilament antibody labels axons cut in cross-section within the white matter, with a secondary labeled with a 488 fluorophore.

The double-label experiment was performed, but no label was seen for the neu-rofilament with the 543 fluorophore (Fig. 14.7a, b, c, Problem experiment). The expected labeling for p38 was seen in the red channel but no neurofilament labeling was seen in the green channel. No 1° antibody control for each antibody (results not shown) confirmed that there was no labeling with mouse anti-p38 alone but samples with rabbit anti-neurofilament did not have labeling.

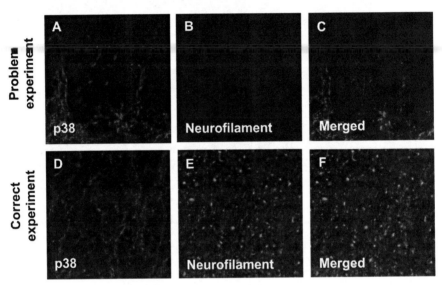

Fig. 14.7 Case No. 4. In this double label experiment, mouse spinal cord was incubated with mouse anti-p38 antibody used at 1:1000 and rabbit anti-neurofilament used at 1:1000 followed by incubation in goat anti-mouse 546 at 1:1000 and goat anti-rabbit 488 fluorophore at 1:1000. (**a**) In the white matter, specific labeling for mouse anti-p38 was high and appropriate. (**b**) The same field was examined for rabbit anti-neurofilament labeling but unexpectedly no labeling was found. (**c**) The merged image from (**a**) and (**b**). (**d**) In the correct experiment the mouse anti-p38 was high and appropriate. (**e**) Now in the same field the rabbit anti-neurofilament labeling was high and appropriate. (**f**) The merged image from (**d**) and (**e**)

Define the Problem

There is no labeling with rabbit anti-neurofilament seen in the double-label experi-ment with mouse anti-p38. The procedure was as follows

Tissue was from a mouse perfused with 4% paraformaldehyde made the same day.

Infiltration was done with 30% sucrose in PBS; sections were cut on a cryostat.

Permeabilization was done with 0.3% Triton X-100 in PBS containing 5% rabbit serum, 10 mg/ml of BSA.

Mouse anti-p38 dilution was 1:1000 and rabbit anti-neurofilament was 1:2000.

Rinses with PBS contained 5% normal rabbit serum, 10 mg/ml of BSA.

Goat anti-mouse with 546 fluorophore 1:1000 and goat anti-rabbit IgG was labeled with 488 fluorophore 1:1000.

Rinses with PBS contained 5% normal rabbit serum, 10 mg/ml of BSA.

Imaging was done with a 40× oil N.A. 1.2 objective on confocal microscope.

Generate Hypotheses

The appearance of this labeling in the experimental and controls suggests a number of potential problems. The following is a list of problems:

(1) Procedural errors
(2) Dilution of the 1° antibodies is incorrect
(3) Dilution of the 2° antibodies is incorrect
(4) 1° antibody is not specific or the antibody has been denatured
(5) 2° antibody does not bind the 1° antibodies
(6) Blocking agents were incorrect

Investigate

The individual hypotheses were evaluated.

(1) Procedural errors – Checks of solutions and dilutions showed no errors.
(2) Dilution of the 1° antibodies is incorrect – The Dilution Matrix had been run and the dilutions of the 1° and 2° antibodies were correct.
(3) Dilution of the 2° antibodies is incorrect – The Dilution Matrix had been run and the dilutions of the 1° and 2° antibodies were correct.
(4) 1° antibody is not specific or the antibody has been denatured – The 1° antibody had been used previously and labeled the neurofilament-containing axons. It is possible that the vial of 1° antibody had been treated improperly and the antibody was denatured. A single antibody experiment is needed to determine whether this vial of 1° antibody still binds to neurofilament.
(5) 2° antibody does not bind the 1° antibodies – The planned repeat of the single antibody experiment in No. 4 will determine whether the 2° antibody still binds to rabbit 1° antibodies.
(6) Blocking agents were incorrect – The blocking serum was 5% rabbit serum and the neurofilament 1° was made in rabbit. Thus, the 2° antibody goat anti-rabbit 488 bound the rabbit IgG in the buffer before the solution was incubated with the tissue section. The 2° antibody goat anti-rabbit 488 bound to the normal

rabbit serum in the incubation solution and no sign of the bound antibody was seen the tissue section.

To better understand this experiment, in the first incubation both 1° antibodies bound to their antigens (Fig 14.8a). With the rabbit serum used as a blocking serum, the goat anti-rabbit 543 fluorophore antibody binds to rabbit IgG of the normal rabbit serum and not the rabbit anti-NF 1° antibody as expected (Fig. 14.8b).

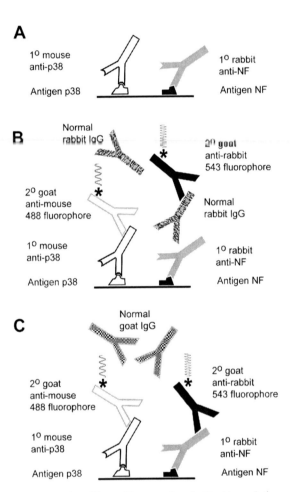

Fig. 14.8 Case No. 4 explanation. The problem was that the blocking solution contained normal rabbit serum that reacted with an antibody. (**a**) Both of the 1° antibodies mouse anti-p38 and rabbit anti-NF (neurofilament) bound to the expected antigens. (**b**) The incubation solution contains normal rabbit serum IgG as a blocking protein. In 2° antibody incubations the goat anti-mouse 488 fluorophore antibody bound to the mouse anti-p38 antibody. However, the goat anti-rabbit 543 fluorophore bound to the rabbit IgG in the blocking serum and not to the rabbit anti-NF, as expected. (**c**) Changing the serum in the incubation solution to normal goat serum allowed both 2° antibodies to find to the appropriate 1° antibodies

When normal goat serum is used in place of normal rabbit serum, the expected double labeling is found (Fig. 14.8c). This case illustrates a problem where antibodies bind in solution and not on the section and it is not possible to see this incorrect binding because the binding in the solution does not give labeling in the tissue but is washed away with the rinses. This lack of evidence of antibody binding (except by seeing no expected label) is difficult to detect and makes the problem solving more difficult. Knowing that antibodies can bind in solution and leave no trace in the tissue is a useful idea to remember when trying to solve immunocytochemistry problems.

The experiment was repeated with 5% goat serum in place of 5% rabbit serum with the same antibodies. As seen in Fig. 14.7d, e, f, the labeling was as expected.

Case No. 5

In cross-sections of the spinal cord in the outer white matter, small axons cut in cross-section are seen with immunocytochemistry for neurofilament as small oval-shaped structures (Fig. 14.9a). In the central gray matter, some randomly oriented axons are seen but cell bodies did not label (Fig. 14.9b, arrows). Neuronal cell bodies

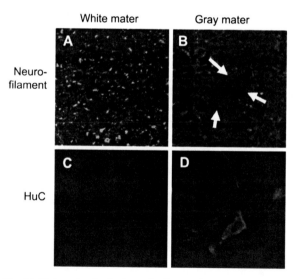

Fig. 14.9 Case No. 5 initial results. In single-antibody experiments, mouse spinal cord was incubated individually with the different 1° antibodies. The spinal cord has a central core called the *gray matter* containing neuronal cell bodies and an outer rim called the *white matter* containing myelinated axons and no neuronal cell bodies. (**a**) Labeling of the white matter with mouse anti neurofilament antibody showed cross-sections of labeled axons as *small dots*. (**b**) Mouse anti-neurofilament antibody-labeled *gray matter* contained randomly oriented axons. (**c**) Mouse anti-HuC-labeled neuronal cell bodies and no labeling was found because no neuronal cell bodies are found in the *white matter*. (**d**) In the *gray matter*, the mouse anti-HuC-labeled the numerous neuronal cell bodies

are seen with immunocytochemistry for HuC in the gray matter (Fig. 14.9d), but no neuronal cell bodies are seen in the white matter (Fig. 14.9c). The images shown in Fig. 14.9 are from single antibody experiments as part of a Dilution Matrix for each 1° antibody.

This fifth and last case is the most difficult and requires thinking outside the box. A double-label indirect immunocytochemistry experiment was performed with these two antibodies. When the white matter was examined normal neurofilament green labeling was seen and unexpected red HuC labeling was seen (Fig. 14.10a, b, *arrows*). The white matter should show green labeling for the axons and no labeling in the red channel (Fig. 14.10c). These images are not consistent with the images obtained from labeling with each of antibodies alone in the white matter (Fig. 14.9a, c). Examining no 1° antibody controls with only mouse anti-HuC showed no labeling was seen in the green or red channels as expected for the white matter (Fig. 14.10c, d). When no 1° antibody control with only rabbit anti-neurofilament was done, the labeling of axons was seen with both green and red fluorescence channels (Fig. 14.10e, f). This duplicates the result found for the incubation with both 1° antibodies (Fig. 14.10a, b).

Define the Problem

Labeling for HuC with the 546 fluorophore shows axons in the white matter that should not be seen in this double-label experiment. The procedure was as follows

> Tissue was from a mouse perfused with 4% paraformaldehyde made the same day.
> Infiltration was done with 30% sucrose in PBS; sections cut on a cryostat.
> Permeabilization was done with 0.3% Triton X-100 in PBS containing 5% normal goat serum, 10 mg/ml of BSA.
> Rabbit anti-neurofilament 1:2000 and mouse anti-HuC 1:500 were used.
> Rinses included PBS containing 5% normal goat serum, 10 mg/ml of BSA.
> Goat anti-rabbit IgG was labeled with 488 fluorophore 1:1000 and goat anti-mouse with 546 fluorophore 1:1000.
> Rinses with PBS contained 5% normal goat serum, 10 mg/ml of BSA.
> Imaging was done with a 63× air N.A. 1.4 objective on a confocal.

Generate Hypotheses

The problem is difficult to define in terms of potential causes. It appears that the anti-HuC antibody labeling was somehow able to bind to axons. Hypothesis generation will include the following, which is a potential list of problems:
(1) Procedural errors
(2) HuC 1° antibody was not specific for antigen
(3) Fluorescent filter sets did not match the fluorophores
(4) Goat anti-mouse 546 fluorophore 2° antibody was not specific
(5) Goat anti-rabbit 488 fluorophore 2° antibody was not specific

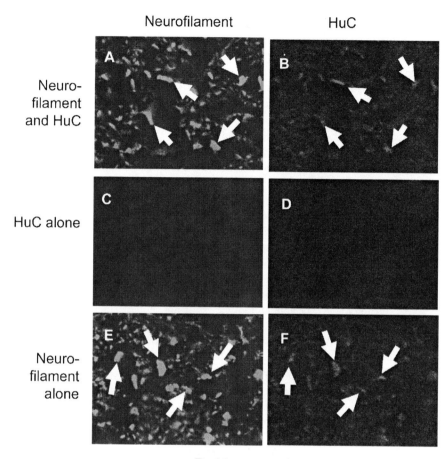

Fig. 14.10 Case No. 5 problem. A double-label antibody experiment with the mouse spinal gave confusing results. The spinal cord *white matter* had only neuronal axons and no neuronal cell bodies. The following micrographs are pairs from double-label experiments showing the *white matter*. (**a**) The labeled showed expected labeling with rabbit anti-neurofilament and goat anti-rabbit 488 fluorophore. (**b**) The same field as a with mouse anti-HuC and goat anti-mouse 546 fluorophore that should have no labeling. Labeling was seen for the same axons (*arrows*) as labeled with anti-neurofilament and 488 fluorophore. (**c**) Incubation without rabbit anti-neurofilament and with mouse anti-HuC followed by both 2° antibodies showed no label in the white matter as expected. (**d**) Same field as (**c**) with no label in the HuC channel. (**e**) Incubation with rabbit anti-neurofilament and without mouse anti-HuC and both 2° antibodies showed the *green labeling* for axons as expected. (**f**) Same field as (**e**) with labeled showed the same unexpected labeling for axons in both channels (*arrows*)

Investigate

The individual hypotheses were evaluated.

(1) Procedural errors – A check of solutions and dilutions showed no errors.

(2) HuC 1° antibody was not specific for antigen – The control results showed that there was no neurofilament labeling with anti-HuC incubations alone in the white matter (Fig. 14.8c, d), suggesting the HuC antibody was not binding to the axons. The control results also showed that with anti-neurofilament alone (Fig. 14.8e, f), the same labeling was seen with both 1° antibodies together (Fig. 14.10a, b). This observation eliminates the mouse anti-HuC 1° antibody as a cause of the problem.

(3) Fluorescent filter set did not match the fluorophore – The filters and lasers were correct. The potential for bleed-through from the lower 488 nm channel into the higher 546 nm channel was checked, but the 546 nm fluorescence did not bleed through from the 488 nm laser.

(4) Goat anti-mouse 546 fluorophore 2° antibody was not specific – This fits the results but experiments with this 2° antibody and a different 1° antibody made in mouse gave the expected results.

(5) Goat anti-rabbit 488 fluorophore 2° antibody was not specific – This also fits the results and the antibody was tried with a different 1° antibody. A new 1° antibody, rabbit anti-glial fibrillary acidic protein antibody, was used with the mouse anti-HuC that surprisingly gave the double labelling of the rabbit anti-glial fibrillary acidic protein antibody (results not shown).

At this point no solution to the problem was found and it was decided to test the idea that the 2° antibodies were responsible. New goat anti-mouse 546 fluorophore and new goat anti-rabbit IgG labeled with 488 fluorophore were obtained and the experiment was repeated. Now the labeling for the mouse anti-HuC was specific for the cell bodies and did not label any axons in the white matter (Fig. 14.11a, b, c, d, e, f). This indicated that one of the previous 2° antibodies were contaminated. An experiment was performed where each 1° was used individually with each 2° antibodies (results not shown) and the problem was that the goat anti-mouse 546 fluorophore was contaminated with an anti-rabbit 546 fluorophore antibody. In labs where multiple users have access to antibodies, contamination of antibodies can be a problem.

To better understand this case, the figure shows the antibody binding (Fig. 14.12). The primary antibodies bound to the correct antigens (Fig. 14.12a). In the original experiment the goat anti-mouse 546 fluorophore was contaminated with an anti-rabbit 546 fluorophore antibody. Thus, rabbit anti-neurofilament antibody bound both the expected goat anti-rabbit 488 fluorophore and the contaminated anti-rabbit 546 fluorophore (Fig. 14.10b). The correct labeling occurred when a new 2° antibody did not have the contaminating antibody (Fig. 14.12c).

High Background Staining

Background labeling is random labeling at lower level than specific labeling and it can take a variety of forms. Most of the times background is diffused and fairly uniform, but sometimes it is particulate and irregular. A list of potential causes includes

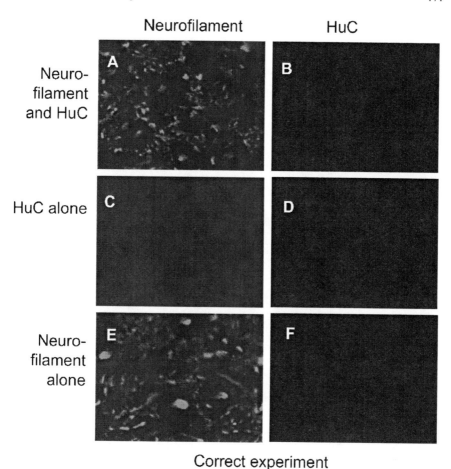

Correct experiment

Fig. 14.11 Case No. 5 correct experiment. The spinal cord *white matter* has only neuronal axons and no neuronal cell bodies. The following micrographs are pairs from double-label experiments showing the *white matter*. (**a**) The cells labeled with rabbit anti-neurofilament and goat anti-rabbit 488 fluorophore showed axons. (**b**) The same filed as a showed no labeling for HuC as expected. (**c**) Incubation with no rabbit anti-neurofilament and with mouse anti-HuC followed by both 2° antibodies showed no label for neurofilament as expected. (**d**) Same field as (**c**) with no label in the HuC channel for axons. (**e**) Incubation with rabbit anti-neurofilament and without mouse anti-HuC and both 2° antibodies showed the labeling for axons. (**f**) Same field as (**e**) with no labeling in the HuC channel

Insufficient rinsing steps before 2° antibody.

The blocking protein or serum was not used at a high enough concentration.

The secondary antibody cross-reacts with endogenous tissue proteins or other endogenous antibody.

The rinses after the 2° antibody were not sufficient.

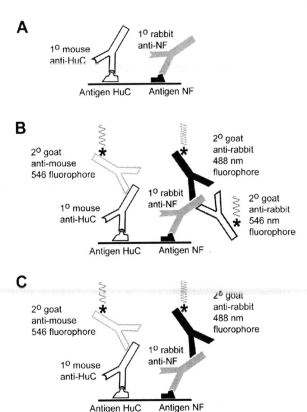

Fig. 14.12 Case No. 5 explanation. The drawing shows how this problem arose. (**a**) Each 1° antibody bound to the correct antigen. (**b**) The 2° antibody incubation showed the goat anti-mouse 546 fluorophore binding the mouse anti-HuC. The goat anti-rabbit 488 fluorophore was the expected 2° antibody, but it was contaminated with goat anti-rabbit 546 fluorophore and both *green* and *red fluorescence* came from the sites with the rabbit anti-NF. (**c**) The correct reagents were used and the HuC antigen was labeled with the mouse anti-HuC followed by the goat anti-mouse 546 fluorophore. The NF antigen was bound by the rabbit anti-NF that was bound by the goat anti-rabbit 488 fluorophore

The sample dried out, attaching reagents on the surface of the section or at the edges.

The tissue might contain Fc receptors or there might be interfering Ig components (aggregates) or there may be naturally occurring antibodies.

Reused antibody solution from previous experiment contains aggregated antibodies.

Particulates exist in the buffer or antibody solution. These can be eliminated by centrifugation or filtration.

Micrograph exposed too long to capture a weak signal.

Incubate with chromogen too long for enzymes.

Troubleshooting Unique to Multiple Primary Antibodies

(1) Use the 1° antibody with the most intense label at highest wavelength fluorochrome
(2) Separate different fluorochromes excitations (and therefore emissions) by at least 60 nm.
(3) Remember that some antibodies incubations can be combined and some need to be separated based on species.
(4) Before trying multiple labels, optimize each antibody in a single antibody experiment.
(5) Multiple labels in wide field fluorescence require band-pass filters and dichroics and multiple labels in confocal sequential image collection.
(6) For all 1° antibodies in an experiment use different species if possible.
(7) For all 2° antibodies in an experiment use the same species and not the same species as any of the 1° antibodies.
(8) Blocking serum for the same species as the 2° antibody.
(9) Use "3rd generation" fluorochromes to minimize fluorescence fading.

Bad Antibodies

Bad antibodies are difficult to troubleshoot and require several experiments to determine they are not the result of procedural or method problems. The best approach is to test the suspected bad antibody in a procedure that works and only replace the one antibody in question.

Bad 1° antibodies from commercial sources are a constant problem for scientists (Couchman, 2009). Many companies do not test their products before they put them on the market. This holds true for different lots or batches of an antibody. When looking for an antibody to a specific protein, the best way is to search the published literature online and determine the source of antibodies what other scientists have successfully used. No one publishes their results with bad 1° antibodies!

Bad 1° Antibodies

- No labeling

   Fixation or detergent specific – There are antibodies that work only with cold methanol fixation or paraformaldehyde.
   Permeabilization – Some antibodies work only with the detergent SDS.
   Antibodies that work in flow cytometry, ELISA or Western but not immunocytochemistry – Generally the method where an antibody works is a function of where it was tested.

Too dilute – Some antibodies might require dilutions of 1:10 and use a lot of reagent

- Too much labeling.

  - Bind to too many epitopes – Western blot will control for this.
  - Non-specific binding – Should be eliminated by using blocking agents.

Bad 2° Antibodies

Buy from well-known companies.

If a problem is suspected, test with known primary antibody in known tissue or cells.

Antibodies become bad if stored improperly.

Chapter 15
Electron Microscopic Immunocytochemistry

Keywords Immunohistochemistry · Antibody labeling · Fluorescence microscopy · Fluorescent immunocytochemistry · Fluorescent immunohistochemistry · Indirect immunocytochemistry · Immunostaining

Contents

Protocol – Pre-embedding Electron Microscopic Immunocytochemistry 175
Introduction 175
Need for Electron Microscopic Immunocytochemistry 176
Pre-embedding Electron Microscopic Immunocytochemistry 178
Postembedding Electron Microscopic Immunocytochemistry 181
Choice of a Method 185
   Advantages and Disadvantages 185
Protocol – Pre-embedding Electron Microscopic Immunocytochemistry 185
Solutions 186
   Stock Solutions to Make Ahead and Store 186
   Solutions Made on the First Day of the Experiment 187
   NPG Silver Enhancement Solution and Silver Lactate 188
   Test Strip 189

Protocol – Pre-embedding Electron Microscopic Immunocytochemistry

Introduction

Electron microscopic immunocytochemistry takes advantage of the increased resolution of the electron microscope to localize label to specific cellular organelles. Optical microscopes limit resolution to the wavelength of light, about 0.2 μm or 200 nm. The electron microscope has theoretical resolution of less than 0.2 nm. Practically for biological samples, resolution is limited to about 2 nm because of the

R.W. Burry, *Immunocytochemistry*, DOI 10.1007/978-1-4419-1304-3_15,
© Springer Science+Business Media, LLC 2010

epoxy resin embedding. Polymers in the epoxy resin scatter the electron beam and limit the resolution of cellular structures.

In the electron microscope, the smallest immunocytochemical label that can be clearly distinguished is 5 nm and easily seen labels need to be 10 nm. To give an idea of relative size, the lipid bilayer of cells is about 5 nm in cross-section thickness and microtubules are 25 nm in diameter. The major problem with applying labels in electron microscopic immunocytochemistry is maintaining the cellular morphology while moving the large labels inside the cells. *As a result there is a constant trade-off between poor morphology, on one hand, and penetration of label into cells, on the other hand.*

To be seen in electron microscopic immunocytochemistry labels must be electron dense. That is, the label must not allow electrons to pass through the label molecule, which thus limits the electron microscopic labels to heavy metals. The most common labels for electron microscopic immunocytochemistry are gold and silver.

It is possible to use HRP as a label for electron microscopic immunocytochemistry because the DAB-developed chromogen generates a reaction product that is easily stained with heavy metals. While HRP can be used for electron microscopic immunocytochemistry, the DAB reaction product does not show high-resolution distribution of the antigen labeled by the 1° antibody. The HRP method has been used to fill specific neurons and trace their dendrites to determine types of synaptic contacts. However, the diffused nature of the reaction product does not allow it to be associated with specific cellular organelles. Therefore, using HRP for electron microscopic immunocytochemistry will not be discussed here.

There are two different ways to perform electron microscopic immunocytochemistry: pre-embedding and postembedding (Stirling, 1990). Pre-embedding electron microscopic immunocytochemistry applies the antibodies and label to samples just after fixation but before embedding in epoxy resin and sectioning. Postembedding electron microscopic immunocytochemistry applies antibodies and label to thin sections made after the samples have been embedded in epoxy resin and sectioned.

Particulate label of electron microscopic immunocytochemistry is frequently quantitated to show clearly which structure is labeled. When looking at the particulate labeling, structures labeled have multiple particles clearly indicating that they contain the antigen. However, single particles are often seen over areas of cells where no antigens are expected, and can be called background labeling. This situation lends itself to quantitation of the particles over structures and adds another layer of complexity to an electron microscopic immunocytochemical experiment. But it should be remembered that, statistically, a single particle is most likely background and cannot be considered to label a structure.

Need for Electron Microscopic Immunocytochemistry

Before deciding to embark on electron microscopic immunocytochemistry, be sure that the needed data cannot be acquired by light microscopic immunocytochemistry.

Shifting from confocal microscopy to electron microscopy requires a completely different method of sample preparation and sectioning, which translates into more time and more money. A first step in preparation is light microscopic immunocytochemistry to determine the correct antibody dilutions, the proper detergent, and the needed blocking agents. Then, as needed, move to electron microscopic immunocytochemistry with the same sample.

Understanding the difference between processing for pre-embedding and postembedding electron microscopic immunocytochemistry provides a basis for selecting the better technique (Table 15.1).

In pre-embedding for electron microscopic immunocytochemistry, the antibodies are all applied to the tissue before embedding in epoxy resin. This immunocytochemical procedure is much like that for light microscopy, followed by epoxy embedding. The label for pre-embedding immunocytochemistry must be small no bigger than an IgG molecule so that it will penetrate into the cells with minimal detergent treatment. The only heavy metal small enough is small gold particles. Undecagold is a core of 11 gold atoms with a single reactive arm for cross-linking (Nanogold®, Hainfeld and Furuya, 1992). However, small gold particles cannot be seen in epoxy section in the electron microscope. The key is to enlarge or enhance these small gold particles with the heavy metal silver, using a photographic developing process called silver enhancement. There are several commercial products for silver enhancement that will increase the size of the gold continuously as long as they are incubated with the cells. This process can give rise to particles

Table 15.1 Conditions for pre-embedding and postembedding electron microscopic immunocytochemistry

Pre-embedding	Postembedding
Fixation	
Trim tissue to blocks	Fixation
Permeabilize	Trim tissue to blocks
1° antibody	
Rinse	
2 ° antibody and small silver label	
Silver enhancement	
Rinse	
Osmium membrane staining	
Dehydration	
Epoxy embedding	Dehydration
Thin sectioning	Epoxy embedding
	Thin sectioning and/or **etching the section**
	1° antibody
	Rinse
	2° antibody and colloidal gold label
	Rinse
Section stain	Section stain
Examine in electron microscope	Examine in electron microscope

Bold unique steps to one procedure

with diameters up to many microns. Before embedding in epoxy resin, the entire immunocytochemical procedure is completed, including silver enhancement, thus giving rise to the name, pre-embedding immunocytochemistry.

In postembedding electron microscopic immunocytochemistry, the antibodies are applied to sections after embedding in epoxy resin. The problem here is that the hydrophobic epoxy resins do not allow antibodies in aqueous solutions to bind to embedded tissue in sections. In this method, aqueous epoxy resins allow antibodies to penetrate into the surface of the thin section. Also, frozen sections from a cryo-ultramicrotome cut on a diamond knife have been used. For processing, thin sections collected on electron microscopic grids are processed section side-down on droplets of antibodies. The label is not limited in size because it will bind to the surface of the thin section and is between 5 and 25 nm in diameter. Colloidal gold is the label of choice and is large enough to be seen directly, without enhancement. Colloidal gold is a colloid of gold atoms surrounded by a protein shell, and as any colloid, it is not a long-term stable reagent.

Each method has its advantages and disadvantages; selecting a method requires a good understanding of the density of the antigen present in the tissue and the potential subcellular organelles where the label is located. An additional consideration is how much subcellular detail needs to be seen in the micrograph to determine where the label is located. One caveat is that these two methods are totally different, with fixation being the only common step. Changing from pre-embedding to postembedding requires starting again to develop the method.

Pre-embedding Electron Microscopic Immunocytochemistry

Pre-embedding electron microscopic immunocytochemistry begins with a tissue block or cultured cells. Samples are fixed with 4% paraformaldehyde with glutaraldehyde ranging from 0 upto 0.3%. The quality of the morphology depends on a higher percentage of glutaraldehyde, but more glutaraldehyde cross-linking prevents antibody penetration. Because pre-embedding method depends on diffusion of antibodies from the exposed or cut surface, cell labeling decreases from the exposed surface and under the best conditions, extends 10 μm into the cells (Fig. 15.1a, b, *arrow* indicates gradient).

The $1^°$ antibody incubations for samples are the same as those used for fluorescent immunocytochemistry, with labeled $2°$ antibodies instead labeled with small gold. The first step is a permeabilization and blocking step, followed by two rinses. The detergent selected will affect the quality of the morphology observed. The strongest detergent is Triton X-100 at 0.3%, although some authors have used upto 1.0% Triton. Triton can make the cell's cytoplasm appear clear or extracted with discontinuous membranes. Generally, Triton is too strong for this method. Either saponin or digitonin, milder detergents, are used at a maximum of 0.3%, which must be included in all incubations and rinses (see Chapter 5, Block and Permeability). Because saponin or digitonin insert into membranes, leaving the membrane integrity

Fig. 15.1 Pre-embedding EM immunocytochemistry. (**a**) Schematic cell with plasma membrane and exposed surface at the *top*. Nucleus is *gray* and to the *left*, several mitochondria, a Golgi apparatus, and multiple vesicles are shown in the cytoplasm. (**b**) Reagents for pre-embedding EM immunocytochemistry diffuse through cells from the exposed surface and leave a concentration gradient with the exposed surface having the highest concentration of reagents

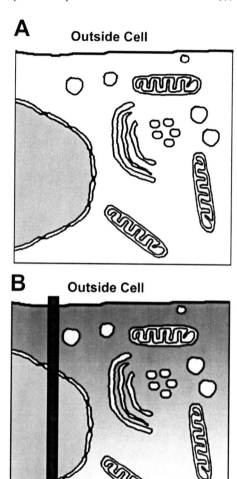

intact, they give the membranes better staining. Following permeabilization, perform two rinses with buffer and the 1° antibody incubation at room temperature with 0.02% sodium azide to prevent bacterial growth.

The 1° antibody incubation can be as short as 6 h or as long as 24 h at a dilution determined by preliminary experiments with fluorescent 2° antibody. All incubations should be done on a rotator to maximize antibody penetration. At the end of the 1° antibody incubation, perform 4–6 rinses in buffer with blocking reagents for 10 min each.

The 2° antibody incubation uses a species-specific antibody labeled with small gold particles (Nanoprobes, or Aurion) (Fig. 15.2a). Using 2° antibodies Fab molecules labeled with small gold gives better penetration because the size of the

Fig. 15.2 Silver
enhancement of small gold.
(**a**) The 1° antibody binds the
antigen and the 2° antibody is
labeled with small gold. (**b**)
The silver enhancement
solution was applied
containing ionic silver and a
chemical developer. (**c**) The
reaction deposited metallic
silver on the gold particle.
(**d**) The reaction continued to
enlarge the particle with
metallic silver until the
enhancement solution is
rinsed off

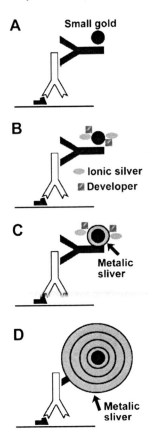

labeled antibody is reduced from using an IgG. The concentrations of the 2° antibody are generally in the range of 1:50. Use the Dilution Matrix (Chapter 10, Single
Antibody Procedure) to determine the correct dilutions for the 1° antibody and the
2° antibody. Also, nanogold has Fab molecules with both fluorescent label and small
gold particles, which allows fluorescent microscopy prior to the silver enhancement
to show the 2° antibody labeling. Prior to silver enhancement, examine the sections
in a fluorescent microscope to confirm that the 1° and 2° antibodies gave the same
labeling pattern as seen in single 1° antibody immunocytochemistry.

Silver enhancement will enlarge small gold particles. The procedure described
below highlights the use of chemicals to mix for each reagent (Burry et al., 1992;
Gilerovitch et al., 1995). There are a number of commercial silver enhancement kits
available, and for kits, the same basic method is applied. The unlabeled 1° antibody
and a small gold-labeled 2° antibody are used (Fig. 15.2a). The silver enhancement
solution contains both a source of ionic silver (silver lactate) and a photographic
developer (*n*-propyl gallate) (Fig. 15.2b). In the presence of small gold, the ionic
silver develops into metallic silver and forms a shell (Fig. 15.2c). This development process can continue and form more metallic silver (Fig. 15.2d) until the silver

enhancement solution is washed away. The length of incubation time determines the size of the final particles. The formation of metallic silver around small gold is not linear. The process starts slowly and reaches an apparent steady state, and this rate will be different from day to day. Thus, silver particles of 20 nm form in 7 min on 1 day and in 20 min the following day.

The silver enhancement solution with its components mixed together will react in a few seconds and turn the solution black or brown. Gum arabic is added to slow the rate of silver enhancement and MES buffer is added to control the pH. It is very important to mix the silverenhancement solution with the reagents in the proper order and exactly as described. The silver enhancement solution described here is sensitive to light, as are most of the commercial solutions, and needs to be mixed and used in a darkroom. Fortunately, the wavelength emitted by a sodium vapor safelight does not affect the solution, but is adequate to perform these procedures.

A most important step here is control the length of time for silver enhancement. Use a test strip that is silver enhanced each time tissue or cells are silver enhanced. The length of time needed to get to a specific size of the silver-enhanced small gold is variable from day to day, and with different batches of silver enhancement solution. The test strip calibrates the incubation time so that intensity, rather than time alone, is used to evaluate development on test spots. To prepare a test strip, use nitrocellulose membranes (also used for western blot, but not PVDF membranes) and spot 1 μl of five different dilutions of the small gold 2° antibody: 1:10, 1:50, 1:100 1:500, and 1:1000. Air-dry the membrane for a several hours and store for no more than 1 week. For testing the silver enhancement solution, use small 4 ml clear disposable tubes and cut the long thin strips to fit in these tubes. Mix 2 ml of enhancement solution under the sodium vapor safelight and rock the tube. At 2 min intervals, view the clear tube under the sodium vapor safelight to determine which spots are visible and how intense are the spots.

For standard electron microscopy, cells and tissue are postfixed in 1% osmium tetroxide for 30 min to stain the membranes. Exposure to osmium tetroxide decreases the size of the silver particles and some might even disappear (Burry et al., 1992). Reducing both the concentration of osmium to 0.1% and processing for 10 min will prevent this loss of silver particles.

The dehydration, embedding, and sectioning of pre-embedding electron microscopic immunocytochemistry blocks follow the standard electron microscopic procedures. However, for section staining, the time is reduced to 2–3 min to preserve the size of the silver particles.

Postembedding Electron Microscopic Immunocytochemistry

Postembedding electron microscopic immunocytochemistry uses tissue or cultured cells that are chemically fixed, sectioned on a diamond knife, and processed for immunocytochemistry. Here, all procedures and reagents, prior to

Fig. 15.3 Postembedding EM immunocytochemistry. (a) Schematic cell with organelles is shown. This method uses ultramicrotome sections of cells with the *lines* indicating a section through the cell. (b) The section is incubated with antibodies and colloidal gold labeling (*arrows*). The colloidal gold is large and remains on the surface of the section. (c) The 1° antibody binds its antigen followed by the colloidal gold-labeled 2° antibody binding the 1° antibody

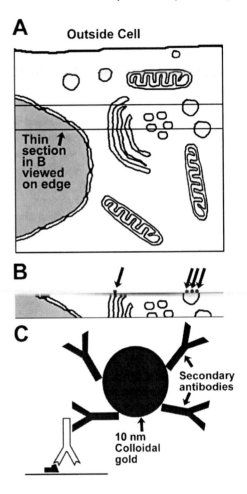

sectioning, must leave the antigens able to bind antibodies in the thin sections. This requires changes to normal electron microscopic procedures that use strong fixatives like glutaraldehyde, postfixation in osmium tetroxide, and hydrophobic embedding resins.

Colloidal gold used as a label is in sizes from 5 to 50 nm. Colloidal gold is easy to identify on sections in the electron microscope because it is uniform in size, uniformly dense, and spherical (Fig. 15.3b). Most commonly, colloidal gold is labeled with protein A that binds the Fc receptor of the primary antibody without the need for a secondary antibody (Ghitescu et al., 1991).

Embedding the tissue with standard hydrophobic epoxy resins will not allow antibody penetration, even to the edge of the tissue section. Two approaches are used to circumvent this problem. First, tissue sections with a standard resin are treated with 10% H_2O_2 to etch the epoxy section on the sections cut surface allowing antibodies to contact proteins in the section. Fixation can be with paraformaldehyde or glutaraldehyde ranging from 0 upto 3%. The normal postfixation staining of membranes with osmium tetroxide, which destroys the antigenicity of the cells, can be used only with some 1° antibodies (Bendayan and Zollinger, 1983).

A second approach to postembedding electron microscopic immunocytochemistry is the use of special "aqueous resins," while these resins are still hydrophobic, they tolerate some aqueous material in the tissue. Examples of these resins are LR White or Lowicryl K4M. Another approach with aqueous embedding is to freeze tissue in buffer very rapidly, and to section with a cryo-ultramicrotome. Cryo-ultramicrotomy is a technique requiring considerable training and cannot be used by the novice.

For aqueous embedding media, fixation is with paraformaldehyde at 4% or glutaraldehyde ranging from 0 upto 3%. Osmium tetroxide cannot be used because it will inhibit the binding of the 1° antibody by destroying the antigenicity of proteins. Instead use of 2% uranyl acetate will allow some membrane staining, even though the lack of membrane staining can create problems in examining the sections in the electron microscope, because there is weak-to-no membrane labeling. For dehydrated of tissue use a graded series of methanol through 100%. A mixture of 50% resin and 50% methanol is used for infiltration followed by resin alone with catalyst for a final infiltration. Hardening of the resin occurs by UV light at temperatures of 4°C or below in the absence of oxygen. LR White can also be infiltrated directly from 70% methanol and be hardened at 50°C. There is the possibility for considerable variability in the infiltration and hardening protocols, depending on the need (Newman and Hobot, 1987).

Hardened resin containing the tissue is sectioned with a diamond knife. Thin sections on nickel grids are then incubated with the antibodies for immunocytochemistry. The protocol here is the same as for a single 1° antibody including separate grids for controls (Chapter 10, Single Primary Antibody). Frequently, Protein A labeled with colloidal gold is used in place of a 2° antibody. The sections contain small amounts of the cytoplasm and organelles from the cells (Fig. 15.4a, b). Using 10 nm colloidal gold as a label allows several 2° antibodies to potentially bind to the 1° antibody (Fig. 15.4c). Because dense nature of the embedding medium, the size of the antibodies and label, they are not able to penetrate into the section (Fig. 15.4c, *arrows*). If frozen sections are used, the penetration will be greater, but still limited by the cellular material and not the embedding material.

A major limitation of the postembedding electron microscopic immunocytochemistry is the lack of penetration of the antibodies and label into the section. For densely concentrated antigens, this is not a limitation. For example insulin in secretory vesicles will have sufficient insulin antigen exposed on the surface of the section to label the vesicle (Fig. 15.4b; *arrows*). However, infrequent cytoplasmic proteins may not have enough antigens exposed on the surface that will be seen over background labeling.

An additional limitation of postembedding electron microscopic immunocytochemistry is its lack of traditional staining for membranes and cellular organelles. Because osmium tetroxide is not used before embedding, membranes can only be seen by negative staining. The use of weaker fixation without high concentration of

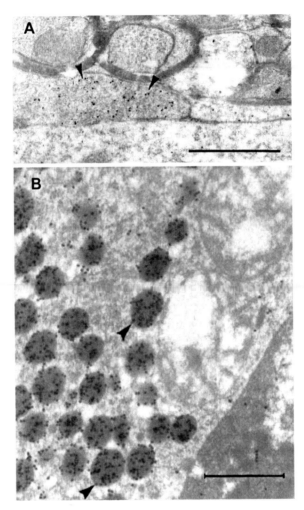

Fig. 15.4 EM immunocytochemistry. (**a**) Pre-embedding labeling of the cerebellum with an antibody to synaptic vesicles. The small *black label* (*arrowheads*) on synaptic terminal indicates labeled vesicles. The plasma membranes, myelin, and mitochondria can be seen; however, some were damaged. Bar is 1 μm. Reproduced with permission Gilerovitch et al., "The Use of Electron Microscopic Immunocytochemistry with Silver-Enhanced 1.4 nm Gold Particles to Localize GAD in the Cerebellar Nuclei" published in the Journal of Histochemistry & Cytochemistry, Vol. 43, No. 3, pp. 337–343 (1995)., 1995. (**b**) Postembedding labeling shows pancreatic cells with an antibody to glicentin. Vesicles are labeled and identified by the dark contents. The membranes of the nuclear envelope and the mitochondria were not seen. Bar is 0.6 μm. Reproduced with permission Varndell et al., "Localization of Glucagon-like Peptide (GLP) Immunoreactants in Human Gut and Pancreas Using Light and Electron Microscopic immunocytochemistry" published in the Journal of Histochemistry & Cytochemistry, Vol. 33, No. 10, pp. 1080–1086 (1985)

glutaraldehyde is needed to allow antigens to be bound by 1° antibodies. The downside of the weaker fixation is that some of the cellular components are washed out during the dehydration steps prior to embedding the tissue.

Choice of a Method

Advantages and Disadvantages

- **Pre-embedding**

    + Good morphology of membranes and organelles
    + Proteins labeled in cells for 15 μm from the exposed surface
    + Demonstrated lower antigen density
    − Silver enhancement, an additional procedure, is required
    − Silver enhancement generates different sizes and shapes particles
    − Multiple antibody experiments are very difficult

- **Postembedding**

    + Easy-to-perform incubations with thin sections on grids
    + Uniform label size and shape
    + Multiple antibody experiments possible
    − Difficult to see membranes and organelles
    − Labeling restricted to high copy proteins available on surface of thin sections
    − Colloidal gold not stable

Protocol – Pre-embedding Electron Microscopic Immunocytochemistry

- **First day**

    (1) In plastic chamber slides with culture cells, fix 4% paraformaldehyde in buffer for 30 min and rinse 4X with PBS.
    (2) Incubate test strip in the *NPG silver enhancement solution* under a sodium vapor safe light until labeling of test strip is similar to that determined previously as acceptable. Note the time required.

- **Second day**

    (3) Permeabilize in PBS+ (0.2% saponin and block 5% normal goat serum and 1% BSA) for 30 min.
    (4) Rinse 2 X PBS+ 5 min each.
    (5) Incubate with mouse 1° antibody for 3 h room temperature.
    (6) Rinse four times PBS+ 5 min each.

(7) Incubate in 2° antibody, GAM 1 nm gold (nanoprobe), 1:50 in PBS+ 1 h at room temperature.

(8) Rinse two times with PBS+ 5 min each and two times in PBS 5 min each.

(9) Fix in 1.6% glutaraldehyde in PBS for 15 min (640 ul/10 ml of 25% Stock glutaraldehyde).

(10) Rinse two times with PBS.

(11) Rinse three times in 50 mM MES buffer with 200 mM sucrose pH = 5.8, over 15 min.

(12) Mix the NPG silver enhancement solution.

(13) Incubate tissue sections in the NPG silver enhancement solution under a sodium vapor safelight until labeling of test strip is similar to that determined previously as acceptable. To be sure of getting the ideal silver enhancement time, it is best to use two or three times for samples to process for the rest of the procedure.

(14) Immediately rinse the samples three times in neutral fixer solution over 5 min under a sodium vapor safelight.

(15) Rinse three times in PBS over 15 min or until gum arabic is gone with microscopic examination.

(16) Incubate for 15 min in 0.1% OsO_4 in rinse buffer (0.1 ml of 4% OsO_4 stock per ml of PBS).

(17) Rinse three times in PBS for 15 min.

- **Third day**

(18) Dehydrate in ethanol (50, 70, 80, 95, 100, 100, and 100%) followed by HPMA and embed in Epon.

(19) Stain thin sections with heavy metals: for cultures 2% uranyl acetate (UA) for 10 min. and lead citrate 3 min. For sections of tissue, use 1% UA for 5 min and lead citrate diluted 1:2 with ddH_2O for 3 min.

Solutions

Stock Solutions to Make Ahead and Store

MES Stock – 0.5 M MES pH = 6.15

MES	1.952 g
Add H_2O	12.0 ml
Adjust pH with 1.0 M NaOH; minimum needed (do not use HCl)	2.0 ml
With H_2O bring volume to	20.0 ml

MES Buffer – 50 mM MES, 200 mM sucrose pH = 5.8

0.5 M MES Stock	1.0 ml
Sucrose	684.0 mg
H$_2$O	7.5 ml
Adjust pH with 1.0 M NaOH (do not use HCl)	
With H$_2$O bring volume to	10.0 ml

Gum Arabic Stock (Sigma G-9752) 50 g/100 ml

Gum Arabic powder	50.0 g
Add volume of H$_2$O	100.0 ml

Combine in a bottle with volume of at least three times the volume of water used. Allow for 2–3 days for the gum arabic to dissolve with gentle agitation. Do not use stir plate. When dissolved, degas with a high-vacuum pump. Use a large flask and control the vacuum because the solution generates extensive bubbles. This step is crucial to the reproducibility of the silver enhancement method. The gum arabic stock can be stored directly in small plastic tubes in a –20°C freezer for no more than 1 year. If you have questions about the suitability of this solution degas and test. Also, test the final solution when new gum arabic is made and when monitoring the silver enhancement during each experiment.

Neutral Fixer Solution 250 mM sodium thiosulfate, 20 mM MES pH = 6.8

Sodium thiosulfate	1.55 g	3.1 g	6.2 g
MES	0.098 g	0.195 g	0.390 g
H$_2$O	22.5 ml	45.0 ml	90.0 ml
Adjust pH with 1.0 M NaOH			
Makes	25.0 ml	50.0 ml	100.0 ml

Solutions Made on the First Day of the Experiment

PBS+ 2.5% normal goat serum, 0.2% saponin, 1% BSA, and 0.02% azide

Normal goat serum	2.5 ml
BSA	1.0 g
Saponin	200.0 mg
Sodium azide	20.0 mg
PBS	97.5 ml
makes	100.0 ml

2° Antibody 1:50

GAM Fab Fluoro Nanogold	40.0 µl
PBS+	1.96 ml
Makes	2.0 ml

NPG Stock; *n*-Propyl gallate 1.0 mg/ml (reduced from 2 mg/ml previously)

NPG (Add to empty tube)	10.0 mg
ETOH	0.25 ml
Mix and dissolve the NPG	
H_2O	9.75 ml
makes	10.0 ml

Silver lactate 3.65 mg/ml (reduced from 7.3 mg/ml previously)

H_2O	5.0 ml
Silver lactate	18.0 mg

Make the day of use. Store in light-tight box, mixing should be done in dimly-lit room lights.

NPG Silver Enhancement Solution and Silver Lactate

Gum arabic stock	5.0 ml	2.5 ml
MES 0.5 M stock pH 6.15 (final 200 mM)	2.0 ml	1.0 ml
NPG Stock	1.5 ml	0.75 ml
Mix for 3 min and add the next solution in at darkroom with a sodium vapor safe light.		
Silver lactate solution	1.5 ml	0.75 ml
Makes	10.0 ml	5.0 ml

Combine in a 4 ml plastic disposable tube and mix by rocking in a light-tight box. After adding the silver lactate, mix for 1 min, and then add to samples.

Test Strip

Use nitrocellulose membrane with serial dilution of 1 nm gold as a test. The dilutions to be dotted at 1 µl on nitrocellulose are f.s., 1:10, 1:50, 1:100, and 1:500. Drying test strips with hair dryer for several minutes. Strips must be made fresh every 3 weeks. After colleting electron microscopic results the correct particle size can be correlated with the intensity of the spots on the test strip. In general, 1:100 dilution gives the best spot size.

Appendix

In the Appendix

Experimental Design Charts

Direct immunocytochemistry
Indirect immunocytochemistry one sample
Indirect immunocytochemistry two sample
Indirect immunocytochemistry block-between
Indirect immunocytochemistry Zenon®
HRP ABC Iimmunocytochemistry

R.W. Burry, *Immunocytochemistry*, DOI 10.1007/978-1-4419-1304-3,
© Springer Science+Business Media, LLC 2010

Experimental design chart
Direct immunocytochemistry

Category	Parameter	Conditions	Antigen No. 1	Antigen No. 2
(1) Sample				
	Source and tissue			
	Fixative			
	Fix application method			
	Embedding			
	Sectioning			
	Incubation chambers/size			
(2) 1° Antibody				
	Antigen			
	Source of 1° antibody			
	Fluorophore			
	Excitation wavelength			
	Emission wavelength			
	Dilution 1° antibody			
	Mixing 1° antibody			
(3) Incubation solutions				
	Buffer			
	Block serum species			
	Detergent			
(4) Controls				
	1° antibody controls			
	Label controls			
(5) Microscope				
	Location			
	Fluorescent filters			
	Lasers			

Experimental design chart
Indirect immunocytochemistry one sample

Category	Parameter	Conditions	Antigen No. 1
(1) Sample			
	Source and tissue		
	Fixative		
	Fix application method		
	Embedding		
	Sectioning		
	Incubation chambers/size		
(2) 1° Antibody			
	Antigen		
	Source of 1° antibody		
	Species of 1° antibody		
	Dilution 1° antibody		
	Mixing 1° antibody		
(3) 2° Antibody			
	Species		
	Source of 2° antibody		
	Fluorophore		
	Excitation wavelength		
	Emission wavelength		
	Dilution 2° antibody		
	Mixing 2° antibody		
(4) Incubation solutions			
	Buffer		
	Block serum species		
	Detergent		
(5) Controls			
	1° antibody controls		
	2° antibody controls		
	Label controls		
(6) Microscope			
	Location		
	Fluorescent filters		
	Lasers		

Experimental design chart

Indirect immunocytochemistry two sample

Category	Parameter	Conditions	Antigen No. 1	Antigen No. 2
(1) Sample				
	Source and tissue			
	Fixative			
	Fix application method			
	Embedding			
	Sectioning			
	Incubation chambers/size			
(2) 1° Antibody				
	Antigen			
	Source of 1° antibody			
	Species of 1° antibody			
	Dilution 1° antibody			
	Mixing 1° antibody			
(3) 2° Antibody				
	Species			
	Source of 2° antibody			
	Fluorophore			
	Excitation wavelength			
	Emission wavelength			
	Dilution 2° antibody			
	Mixing 2° antibody			
(4) Incubation solutions				
	Buffer			
	Block serum species			
	Detergent			
(5) Controls				
	1° antibody controls			
	2° antibody controls			
	Label controls			
(6) Microscope				
	Location			
	Fluorescent filters			
	Lasers			

Experimental design chart

Indirect immunocytochemistry block-between

Category	Parameter	Conditions	Antigen No. 1	Antigen No. 2
(1) Sample				
	Source and tissue			
	Fixative			
	Fix application method			
	Embedding			
	Sectioning			
	Incubation chambers/size			
(2) 1° Antibody				
	Antigen			
	Source of 1° antibody			
	Species of 1° antibody			
	Dilution 1° antibody			
	Mixing 1° antibody			
(3) 2° Antibody				
	Species			
	Source of 2° antibody			
	Fluorophore			
	Excitation wavelength			
	Emission wavelength			
	Dilution 2° antibody			
	Mixing 2° antibody			
(4) Block-between				
	Dilution normal 1° species serum			
	Dilution of anti-species 1° Fab			
(5) Incubation solutions				
	Buffer			
	Block serum species			
	Detergent			
(6) Controls				
	1° antibody controls			
	2° antibody controls			
	Label controls			
(7) Microscope				
	Location			
	Fluorescent filters			
	Lasers			

Experimental design chart

Indirect Immunocytochemistry Zenon®

Category	Parameter	Conditions	Antigen No. 1	Antigen No. 2
(1) Sample				
	Source and tissue			
	Fixative			
	Fix application method			
	Embedding			
	Sectioning			
	Incubation chambers/size			
(2) 1° Antibody				
	Antigen			
	Source of 1° antibody			
	Species of 1° antibody			
	1° Antibody subclass			
	Diluting 1° antibody X 2			
	Mixing antibody			
(3) 1° Antibody labeling				
	Source of Fab			
	Fluorophore			
	Excitation wavelength			
	Emission wavelength			
	Amount Fab Zenon component A			
	Amount normal mouse serum			
(4) Incubation solutions				
	Buffer			
	Block serum species			
	Detergent			
(5) Controls				
	1° antibody controls			
	2° antibody controls			
	Label controls			
(6) Microscope				
	Location			
	Fluorescent filters			
	Lasers			

Experimental design chart
HRP ABC immunocytochemistry

Category	Parameter	Conditions	Antigen No. 1
(1) Sample			
	Source and tissue		
	Fixative		
	Fix application method		
	Embedding		
	Sectioning		
	Incubation chambers/size		
(2) 1° Antibody			
	Antigen		
	Source of 1° antibody		
	Species of 1° antibody		
	Dilution 1° antibody		
	Mixing 1° antibody		
(3) 2° Antibody			
	Species antibody 2° labeled with biotin		
	Source antibody 2° labeled with biotin		
	Diluting antibody 2° labeled with biotin		
	Mixing antibody 2° labeled with biotin		
	Source of ABC reagent		
	Mixing ABC		
	HRP chromogen		
	Mixing chromogen		
(4) Incubation solutions			
	Buffer		
	Block serum species		
	Detergent		
(5) Controls			
	1° antibody controls		
	2 antibody controls		
	Label controls		
(6) Microscope			
	Location		
	Fluorescent filters		
	Lasers		

References

Angelov DN, Walther M, Streppel M, Guntinas-Lichius O, Neiss WF (1998) The cerebral perivascular cells. Adv Anat Embryol Cell Biol 147: 1–87.

Baschong W, Suetterlin R, Laeng RH (2001) Control of autofluorescence of archival formaldehyde-fixed, paraffin-embedded tissue in confocal laser scanning microscopy (CLSM). J Histochem Cytochem 49: 1565–1572.

Beisker W, Dolbeare F, Gray JW (1987) An improved immunocytochemical procedure for high-sensitivity detection of incorporated bromodeoxyuridine. Cytometry 8: 235–239.

Bendayan M, Zollinger M (1983) Ultrastructural localization of antigenic sites on osmium-fixed tissues applying the protein A-gold technique. J Histochem Cytochem 31: 101–109.

Billinton N, Knight AW (2001) Seeing the wood through the trees: a review of techniques for distinguishing green fluorescent protein from endogenous autofluorescence. Anal Biochem 291: 175–197.

Burry RW, Vandre DD, Hayes DM (1992) Silver enhancement of gold antibody probes in pre-embedding electron microscopic immunocytochemistry. J Histochem Cytochem 40: 1849–1856.

Burry RW (1995) Pre-embedding immunocytochemistry with silver-enhanced small gold particles. In: Hayat MA (ed.) Immunogold-silver staining: principles, methods and applications. Boca Raton, FL: CRC Press, pp 217–230.

Burry RW (2000) Specificity controls for immunocytochemical methods. J Histochem Cytochem 48: 163–166.

Clancy B, Cauller LJ (1998) Reduction of background autofluorescence in brain sections following immersion in sodium borohydride. J Neurosci Methods 83: 97–102.

Coons AH, Creech HJ, Jones RN, Berliner E (1942) The demonstration of pneumococcal antigen in tissues by the use of fluorescent antibody. J Immunol 45: 159–170.

Couchman JR (2009) Commercial antibodies: the good, bad, and really ugly. J Histochem Cytochem 57: 7–8.

Ghitescu L, Galis Z, Bendayan M (1991) Protein AG-gold complex: an alternative probe in immunocytochemistry. J Histochem Cytochem 39: 1057–1065.

Gilerovitch HG, Bishop GA, King JS, Burry RW (1995) The use of electron microscopic immunocytochemistry with silver-enhanced 1.4-nm gold particles to localize GAD in the cerebellar nuclei. J Histochem Cytochem 43: 337–343.

Glazer AN, Stryer L (1983) Fluorescent tandem phycobiliprotein conjugates. Emission wavelength shifting by energy transfer. Biophys J 43: 383–386.

Good NE, Winget GD, Winter W, Connolly TN, Izawa S, Singh RM (1966) Hydrogen ion buffers for biological research. Biochemistry 5: 467–477.

Hainfeld JF, Furuya FR (1992) A 1.4-nm gold cluster covalently attached to antibodies improves immunolabeling. J Histochem Cytochem 40: 177–184.

Heffer-Lauc M, Viljetic B, Vajn K, Schnaar RL, Lauc G (2007) Effects of detergents on the redistribution of gangliosides and GPI-anchored proteins in brain tissue sections. J Histochem Cytochem 55: 805–812.

Herzenberg LA, Parks D, Sahaf B, Perez O, Roederer M, Herzenberg LA (2002) The history and future of the fluorescence activated cell sorter and flow cytometry: a view from Stanford. Clin Chem 48: 1819–1827.

Hoffman GE, Le WW, Sita LV (2008) The importance of titrating antibodies for immunocyto-chemical methods. Curr Protoc Neurosci Chapter 2: Unit.

Houser CR, Barber RP, Crawford GD, Matthews DA, Phelps PE, Salvaterra PM, Vaughn JE (1984) Species-specific second antibodies reduce spurious staining in immunocytochemistry. J Histochem Cytochem 32: 395–402.

Hsu SM, Raine L, Fanger H (1981) Use of avidin-biotin-peroxidase complex (ABC) in immunoperoxidase techniques: a comparison between ABC and unlabeled antibody (PAP) procedures. J Histochem Cytochem 29: 577–580.

Kuhlmann WD, Peschke P (1986) Glucose oxidase as label in histological immunoassays with enzyme-amplification in a two-step technique: coimmobilized horseradish peroxidase as secondary system enzyme for chromogen oxidation. Histochemistry 85: 13–17.

Larsson L-I (1988) Immunocytochemistry: theory and practice. Boca Raton, FL: CRC Press.

Lewis Carl SA, Gillete-Ferguson I, Ferguson DG (1993) An indirect immunofluorescence pro-cedure for staining the same cryosection with two mouse monoclonal primary antibodies. J Histochem Cytochem 41: 1273–1278.

McGhee JD, von Hippel PH (1975) Formaldehyde as a probe of DNA structure. I. Reaction with exocyclic amino groups of DNA bases. Biochemistry 14: 1281–1296.

McLean IW, Nakane PK (1974) Periodate-lysine-paraformaldehyde fixative. A new fixation for immunoelectron microscopy. J Histochem Cytochem 22: 1077–1083.

Nakane PK, Pierce GB, Jr. (1966) Enzyme-labeled antibodies: preparation and application for the localization of antigens. J Histochem Cytochem 14: 929–931.

Neumann M, Gabel D (2002) Simple method for reduction of autofluorescence in fluorescence microscopy. J Histochem Cytochem 50: 437–439.

Newman GR, Hobot JA (1987) Modern acrylics for post-embedding immunostaining techniques. J Histochem Cytochem 35: 971–981.

North AJ (2006) Seeing is believing? A beginners' guide to practical pitfalls in image acquisition. J Cell Biol 172: 9–18.

Pearson H (2007) The good, the bad and the ugly. Nature 447: 138–140.

Polak JM, Van Noorden S (2003) Introduction to Immunocytochemistry. Oxford, UK: BIOS Scientific Publishers Limited.

Renshaw S (2007) Immunohistochemistry: Methods Express Series. Oxford, UK: Scion Publishing Ltd.

Riggs JL, Seiwald RJ, Burckhalter JH, Downs CM, Metcalf TG (1958) Isothiocyanate compounds as fluorescent labeling agents for immune serum. Amer J Path 34: 1081.

Rosene DL, Roy NJ, Davis BJ (1986) A cryoprotection method that facilitates cutting frozen sec-tions of whole monkey brains for histological and histochemical processing without freezing artifact. J Histochem Cytochem 34:1301–1315.

Rossner M, Yamada KM (2004) What's in a picture? The temptation of image manipulation. J Cell Biol 166: 11–15.

Sabatini DD, Bensch K, Barrnett RJ (1963) Cytochemistry and electron microscopy. The preser-vation of cellular ultrastructure and enzymatic activity by aldehyde fixation. J Cell Biol 17: 19–58.

Schmiedeberg L, Skene P, Deaton A, Bird A (2009) A temporal threshold for formaldehyde crosslinking and fixation. PLoS ONE 4(2): e4636.

Schnell SA, Staines WA, Wessendorf MW (1999) Reduction of lipofuscin-like autofluorescence in fluorescently labeled tissue. J Histochem Cytochem 47: 719–730.

Shu SY, Ju G, Fan LZ (1988) The glucose oxidase-DAB-nickel method in peroxidase histochemistry of the nervous system. Neurosci Lett 85: 169–171.

Solomon MJ, Varshavsky A (1985) Formaldehyde-mediated DNA-protein crosslinking: a probe for in vivo chromatin structures. Proc Natl Acad Sci U S A 82: 6470–6474.

Sternberger LA, Hardy PH, Jr., Cuculis JJ, Meyer HG (1970) The unlabeled antibody enzyme method of immunohistochemistry: preparation and properties of soluble antigen-antibody complex (horseradish peroxidase-antihorseradish peroxidase) and its use in identification of spirochetes. J Histochem Cytochem 18: 315–333.

Stirling JW (1990) Immuno- and affinity probes for electron microscopy: a review of labeling and preparation techniques. J Histochem Cytochem 38: 145–157.

Stirling JW (1993) Controls for immunogold labeling. J Histochem Cytochem 41: 1869–1870.

Swaab DF, Pool CW, Van Leeuwen FW (1977) Can specificity ever be proved in immunocytochemical staining. J Histochem Cytochem 25: 388–391.

van der Loos CM (2008) Multiple immunoenzyme staining: methods and visualizations for the observation with spectral imaging. J Histochem Cytochem 56: 313–328.

Varndell IM, Bishop AE, Sikri KL, Uttenthal LO, Bloom SR, Polak JM (1985) Localization of glucagon-like peptide (GLP) immunoreactants in human gut and pancreas using light and electron microscopic immunocytochemistry. J Histochem Cytochem 33: 1080–1086.

Wolf WP, Weis M, von Scheidt W (2001) Endothelin immunocytochemistry: indications of false-positive labeling patterns and non-detectable antigen concentrations. Histochem Cell Biol 116: 411–426.

Glossary

Chapter 1

Epitope retrieval or antigen retrieval – done on formalin-fixed paraffin sections involves heating the sections in buffer with either an acid or base to allow the antibody to bind the epitope.

Immunocytochemistry – the use of antibodies in animal research with cells and tissues fixed in paraformaldehyde and sectioned on a cryostat.

Immunohistochemistry – the use of antibodies to study paraffin sections, mainly from human tissue.

Individual studies – information about individual cells as to their location, their relationship to other cells, and their individual complement of molecules.

Population studies – many cells pooled for analysis; studies are duplicated and results are presented with statistically reliable average values with standard errors.

Chapter 2

Affinity purification – an antigen in a column binds antibodies of interest that are then eluted from a column.

Antigen – a protein, peptide, or molecule used to cause an immune response in an animal containing one or more epitopes.

Ascites fluid – generated in the peritoneal cavity (abdominal cavity) of a mouse as a response to injected hybridoma cells; contains concentrated monoclonal antibody.

Constant region – the species-specific sequences and the Fc-binding domain of an IgG.

Epitope – a specific amino acid sequence on a denatured peptide or a several sequences on the surface of a folded protein that generate a clone of antibodies; more than one epitope can exist on an antigen.

Hybridoma cells – cultured cells producing an antibody to a single epitope.

IgG – the antibody isotype preferred for immunocytochemistry

Ig isotypes or classes of antibodies (IgA, IgD, IgE, IgG, IgM) generated under different circumstances.

Immunoglobulin (Ig) – proteins or antibodies generated by animals in response to a challenge.

Monoclonal antibodies – a single antibody from one clone of B-cells to a single epitope on the antigen; these immortal cultured cells can be used to generate antibodies forever, mainly from mouse and rabbit.

Polyclonal antibodies – multiple clones of antibodies produced to different epitopes on the antigen; produced in a single animal from its serum.

Primary antibody (1° antibody) – in immunocytochemistry, the antibody that binds directly to the antigen.

Purified Ig – separated from other serum proteins and all other IgGs; includes the antibody of interest

Supernatant – tissue culture media from hybridoma cells that produce a monoclonal antibody

Variable region – the portion of an antibody with the unique configuration that specifically binds the epitope with the fraction antigen binding (Fab) region.

Whole serum – derived from an animal that generates polyclonal antibodies.

Chapter 3

Acrolein (C_3H_4O) – an exceedingly fast penetrating fixative that contains both an aldehyde and a double bond. Although an excellent fixative, acrolein is highly poisonous, causes severe irritation to exposed skin, is extremely flammable, and is a mild carcinogen.

Bouin's – a denaturing fixative consisting of 70% saturated picric acid, 10% formalin, and 5% acetic acid. This fixative is mainly used for paraffin material and is both highly explosive and carcinogenic.

Cross-linking fixation – reactive groups bind proteins and lipids in cells and holds them in the same position as they were in living cell; the fixative of choice for immunocytochemistry.

Denaturing fixation – destruction of the molecular structure of cellular molecules; for proteins, denaturing breaks the 3D protein structure.

Drop-in fixation – removal of tissue from a live animal followed by placing in ice-cold fixative solution; to increase access of the fixative solution to the tissue, the tissue is cut with a sharp scalpel into blocks.

Fixation – the stabilization or preservation of cells and tissues as close to life-like as possible.

Formaldehyde (CH_2O) – a cross-linking fixative; the best fixative for light microscopic immunocytochemistry.

Formalin – commercial fixative produced by oxidation of methanol and contains 37% formaldehyde and impurities including 14% methanol, small amounts of formic acid, other aldehydes, and ketones; used in human clinical pathology samples.

Glutaraldehyde ($C_5H_8O_2$) – a cross-linking chemical fixative that is too effective at cross-linking because it inhibits the diffusion of antibodies into cells and tissues in immunocytochemistry

Good's Buffers – organic buffers selected by Dr. Norman Good because of their favorable characteristics in biology and biochemistry

Hypertonic solution – when outside the cell water flows out of the cell and into the extracellular space, causing the cell to shrink because the molecules are more concentrated (has less water) than that inside the cell

Hypotonic solution – when outside the cell water flows into the cell and the cell swells because there are less molecules (has more water) than inside the cell.

Isotonic solution – allows movement of water into and out of the cell; there is no net change in the size of the cell.

Paraformaldehyde – a powder of polymerized formaldehyde that is made into pure formaldehyde with heat at high pH; does not contain any of the impurities of formalin; used in biomedical animal research.

Periodate-lysine-paraformaldehyde (PLP) – fixative used to increase cross-linking of molecules by oxidizing sugars attached to lipids and proteins generating aldehydes; binds lysine and is further are cross-linked by formaldehyde.

Phosphate buffer – best vehicle for immunocytochemistry.

Vascular perfusion – uses an animal's blood vessels including capillaries to carry a fixative to the cells quickly before cells can undergo necrotic cell death from lack of nutrients or oxygen.

Vehicle – solution or buffer solution used to maintain pH and tonicity of fixatives.

Zenker's – a heavy-metal, denaturing fixative with mercuric chloride and potassium dichromate; is both highly corrosive and carcinogenic.

Chapter 4

Cryostat – a microtome in a freezer used to cut sections of frozen, infiltrated tissue.

Floating section immunocytochemistry – performed on free-floating, thick section (25–100 mm) cut on a Vibratome or freezing microtome with antibodies penetrating deeper because of the greater movement of the section floating both sides of the section will be exposed to the antibodies.

Freezing microtome – used for cutting 15- to 50-μm thick sections of fixed tissue after infiltrated with 20% sucrose and quickly frozen with liquid CO_2; sections are for floating incubations.

Fresh frozen tissue – samples with no fixation are frozen and sectioned; gives distorted the morphology because no cross linking fixative and components of the tissue can wash out of the section.

Hydrophobic Pen – sections on microscope slides are outlined with hydrophobic material in the form of a box from a PAP-Pen or ImmunoPen to keep the solutions confined during incubations.

Infiltration – processing of tissue blocks with by in 20% sucrose in buffer with agitation overnight at 4°C (cold room) to cryoprotected the tissue for freezing.

Isopentane – a chemical used for rapid freezing tissue blocks at –160°C.

Microwaves – be used for ultra-rapid processing of tissue and speeding incubation times.

O.C.T. (optimum cutting temperature) – is an embedding liquid used to surround tissue blocks for freezing on a cryostat chuck.

Snap freezing – widely used term to describe freezing tissue for immunocytochemistry; the term comes from food preparation industry and for biomedical sciences it has no single definition and can mean freezing in liquid nitrogen, on dry ice, or in isopentane.

Vibratome – a device used on delicate, fixed tissue before sectioning; a vibrating razor blade cuts large pieces of tissue into sections 25–500 μm for freezing or processing.

Chapter 5

Bovine serum albumen (BSA) – the most common blocking agent; works because albumen is non-antigenic and will not bind to the Fab end of an antibody.

Ionic detergents – contain highly charged groups and apolar groups and are very good solubilizing agents; examples are SDS, Deoxycholate, and CHAPS.

Ionic detergents – have highly charged groups and are very good solubilizing agents example is SDS.

Nonionic detergents – have weaker groups for hydrogen bonds and are able to solubilize membranes but are less denaturing example is Triton X-100.

Nonionic detergents – weaker groups for hydrogen bonds, apolar groups and are able to solubilize membranes; examples are, Triton X-100, and Tween 20.

Nonspecific antibody binding – occurs when an antibody binds by a mechanism other than by epitope binding.

Normal serum – from an animal that has not been immunized or exposed to antigens; this serum contains IgGs to only expected antigens.

Triton X-100 – a nonionic detergent commonly used in permeabilizing membranes of fixed tissue for immunocytochemistry.

Chapter 6

Emission – a photon is generated when a relaxed singlet electron falls back to its ground state; its energy is converted into a photon with a lower energy and higher wavelength than the excitation electron.

Excitation – absorption of a photon by a fluorophore is accomplished when a photon of a specific wavelength strikes a fluorophore knocking, an orbital ground state electron to that of an unstable excited singlet electron.

Fluorescent molecules (fluorophore) – molecules able to absorb one wavelength of light and emit a higher wavelength of light; used as labels for antibodies in immunocytochemistry.

Horseradish peroxidase (HRP) – the enzyme most commonly used to label antibodies; its reaction product, called a chromogen, is seen in bright-field microscopes.

Photobleaching – the oxidation of the fluorescence molecule so that it structure is changed with the loss of ability to generate emission photons.

Quantum dots – heavy-metal nanocrystals that fluoresce when exposed to UV light, a major advantage of Qdots is that they do not photobleach; however, their large size makes them difficult to use for immunocytochemistry.

Quantum yield – a measure of the efficiency of a fluorophore to convert excitation photons to emission photons.

Quenching – when the energy of the excited singlet is transferred to another very close molecule, with no emission photon from the excited fluorophore.

Stokes Shift – the difference in the wavelengths of the excitation and emission photon foa fluorophore.

Chapter 7

Avidin – an egg white protein or as a streptavidin from *Streptomyces avidinii* bacteria that has four extremely high-affinity-binding sites for biotin.

Avidin–biotin complex (ABC) – a reagent made from avidin, biotin, and HRP, which dramatically increases detection sensitivity for immunocytochemistry.

Biotin – a very small molecule, also known as vitamin B7, that binds avidin with extremely high affinity; each biotin has just one binding domain for avidin.

Direct avidin–biotin method – a three-step immunocytochemical procedure that uses biotin-labeled 2° antibody to bind the 1° antibody and in a third step, labeled avidin is bound to the biotin of a 2° antibody.

Direct immunocytochemistry – a method where the label is bound to the 1° antibody, which then binds to the antigen in the cells.

Indirect avidin–biotin method – a four-step method with two additional incubation steps after the 1° antibody; the 2° antibody with biotin is followed by incubation with an unlabeled avidin and finally a labeled biotin.

Indirect immunocytochemistry – a method where the labeling comes from a labeled secondary (2°) antibody binding the 1° antibody, in a second incubation step.

Secondary (2°) antibody – a species-specific antibody that binds to the constant end of other antibodies in immunocytochemistry; frequently, it is labeled with a marker seen in the microscope.

Tyramide signal amplification (TSA) – a powerful amplification method that uses labeled tyramide, which is activated by HRP and binds tyrosine amino acids on adjacent proteins.

Chapter 8

Adsorption controls – uses isolated antigens to bind the 1° antibody inhibiting its ability to bind to antigens in the tissue during incubation.

Autofluorescence – fluorescence observed in fixed cells without incubation in any fluorescent compound.

Controls – in immunocytochemistry it is essential to rule out the possibility of a nonspecific labeling being identified as specific labeling with additional steps.

Endogenous peroxidase activity – caused by naturally occurring peroxidase in cells that will generate a chromogen reaction product in the absence of HRP.

Label control – used to show labeling results from the labels attached to the antibodies.

Primary (1°) antibody control – used to show specific binding of the 1° antibody to its antigen.

Secondary (2°) antibody control – used to show specific 2° antibody to the 1° antibody.

Chapter 9

Detection resolution – a measure of the degree to which the label localizes to the 1° antibody site in the cell.

Detection sensitivity – a measure of the intensity of label in the microscope following the immunocytochemical procedure.

Experimental Design Chart – a guide to the selection of antibodies, incubation conditions, and solutions in planning an immunocytochemical experiment.

Chapter 10

Antibody Dilution Matrix – a method to determine the optimal dilution for both the 1° antibody and the 2° antibody

Aqueous mounting media – a solution miscible with buffers, does not need dehydration, and is used with fluorescence immunocytochemistry.

Chapter 11

Multiple 1° antibodies from different species – a method with two 1° antibodies made in different species are combined in one incubation and the appropriate 2° antibodies are combine in one incubation.

Chapter 12

Multiple 1° antibodies made in the same species block between – a method where two 1° antibodies made in same species are used two separate incubations for 1° and 2° antibody separated by a blocking step with normal serum and Fab fragments.

Multiple 1° antibodies made in the same species with Zenon – a method where different 1° antibodies are labeled with species specific Fab fragments separately and then combining them for a single incubation step.

Chapter 13

Antifade or antibleach agents – reducing compounds that lower the amount of free oxygen in the mounting medium and reduce the chance of photobleaching.

Band pass fluorescent filter set – transmits only a narrow band of excitation wavelength to pass and be observed.

Bleed-through – in images with two fluorophores the detection of emission from the fluorophore with lower emission in the filter set of a higher wavelength fluorophore.

Dichroic mirror – a device that reflects excitation wavelengths to the sample and emitted wavelengths pass through the dichroic to be viewed; the dichroic, sometimes called the "beam splitter".

Emission filter – a device above the dichroic and before the eyepiece that allows only photons emitted from the fluorophore to be seen.

Excitation filter – a device near the light source that eliminates photons except those needed to excite the fluorophore.

Filter set – a device that includes three elements: the excitation filter (near the light source) that eliminates photons except those needed to excite the fluorophore, the emission filter (above the dichroic) that allows only photons emitted from the fluorophore to be seen and the dichroic mirror that reflects excitation photons and allow passage of emission photons.

Fluorescence microscope – a microscope that illuminates the sample from above with emission wavelengths by a dichroic mirror and excitation wavelengths are detected.

Fluorophore – a fluorescent molecule that is excited by a wavelength of light and emits a higher wavelength of light; used in fluorescent immunocytochemistry

Long pass fluorescent filter set – transmits all excitation photons above a cutoff wavelength; this filter set allows maximum detection of a fluorophore emission.

Photobleaching – the reduced output of a fluorophore due to irreversible damage to the molecule in the presence of molecular oxygen; permanent loss of the fluorophore molecule is a significant problem for immunocytochemistry.

Quenching – the reduced output of fluorophore because its emission energy is absorbed by adjacent molecules such as oxidizing agents, salts, heavy metals, or other fluorochromes, and no emission photon is generated.

Chapter 15

Colloidal gold – labeling for postembedding electron microscopic immunocyto-chemistry uses a colloids of gold with sizes from 5 to 50 nm attached to label secondary antibodies or protein A.

Electron microscopic immunocytochemistry – uses antibodies to label structures with 5–50 nm gold particles that are examined with an electron microscope.

Postembedding electron microscopic immunocytochemistry – a method that uses antibodies applied to sections after embedding in epoxy resin and sectioning with an ultramicrotome.

Pre-embedding electron microscopic immunocytochemistry – a method that uses antibodies for incubation in the tissue before the embedding in epoxy resin.

Silver enhancement – used in pre-embedding electron microscopic immuno-cytochemistry to enlarge small gold particles to a size seen in the electron microscope.

Small gold particles – used in pre-embedding electron microscopic immunocyto-chemistry; silver enhanced to be seen in the electron microscope.

Index

Note: The letters 'f' and 't' following locators refer to figures and tables respectively. Locators in **bold** refer to definitions of the respective key terms.

A

ABC, *see* Avidin-Biotin Complex (ABC)
Acrolein (C_3H_4O), **21**, 21f
Adsorption controls, **82**, 82f, 83, 83f
  antigen/antibody ratio, 82
  negative absorption control, 82
  positive absorption control, 82
    antibody–protein complex, 83
    protein–protein binding, 83
  purified antigens, 83
  titration of antigen, 82
Affinity purification, **13**
"Aging pigment," 86
"Ah-ha" moment, 152
Aldehyde groups, 86
Allophycocyanin (APC), 60
Angelov, D. N., 62–63
Antibleaching agents, 106
Antibodies
  antibody generation
    by B-cells, 10
  antibody molecules, 8–10
    isotypes/classes (*IgG* antibody
      isotype), 8t
  handling and storage
    storage freezer (–20°C),
      recommended, 16
    storage refrigerator 4°C,
      recommended, 16
  immunocytochemistry, key reagent of, 7
    properties of antibodies, 7
  preparation of
    forms and purity, 13
    monoclonal antibodies, 11–13, 11f
    polyclonal antibody, 10f, 10–11
  primary (1) antibodies, 15
  product information list
    antibody type or host, 14
    catalogue information, 14

    description or background, 14
    packaging, product, or purification, 14
    protocols, 15
    source of antigen, 14
    species reactivity, 15
    specificity, 14
    uses or application, 14–15
  rabbit anti-tubulin antibody, 13
1 antibody controls, 80–83
  absorption controls, 82–83, 83f
  adsorption controls, 82, 82f
  1 antibody-binding antigens, 81f
  epitope and antigen, distinction, 80
  recommendations, 83–84
  specificity control methods, 80, 82f
    immunoblot (western blot), 80–81
    immunocytochemical localization,
      comparison of, 81
    tissue sections, comparibility, 80
  specificity/non-specificity, 80
2 antibody controls, 84–85
  blocking agents, 84
  detection method, 84
  human IgG, 85
  mouse 1 antibodies, 85
  multiple 1 and 2 antibodies, 85
  multiple antibody methods, 85
  negative control or technique control, 84
  nonspecific binding, 84
Antibody detection methods, 93
Antibody dilution matrix, **101**–102, 108–109
  numerical evaluation of labels using
    micrographs, 102t, 103f
Antibody labeling, 2, 79, 89
  for bright field microscopy
    enzymes, use of, 56
    fluorescent labels, use of, 56
    particulates (enhanced gold or silver),
      use of, 56

Antibody labeling (*cont.*)
  dyes used
    eosin, 55
    hematoxylin, 55
  enzyme theory
    enzyme substrates, 61–63
  with a fluorescence compound (Albert
      Coons), 55
  fluorescence theory, 56–58
  fluorescent/enzyme methods for
      immunocytochemistry
    advantages/disadvantages, 64
  fluorescent labels, generations of, 58–59
  immunocytochemistry fluorophores and
      flow cytometry
    choosing fluorochromes, guidelines
        for, 61
  particulate label, 63–64
Antibody penetration
  blocking agents, effects on, 51–52, 52f
  depth of penetration, criteria for
      achieving, 49
  permeabilizing tissue/cells for, 49–51
    hydrophilic/hydrophobic face, 49
    immunocytochemistry, use of
        detergents in, 50f
    membrane permeabilization, 49f
    membranes as lipid bilayers, 49
    organic solvents as denaturing fixatives,
        use, 50
    transmembrane proteins, role in, 49
Antifade agents, **146**
Antigen, **9–10**
  antibody generation by, 9f
Anti-mouse Fab fragment, 121f, 122, 125–127
APC, *see* Allophycocyanin (APC)
Application label and method, 89–93
  application method, 89–90
  choosing application, 89–93
  complete planning strategy, 89
    exploration and selection, 89
    preparing reagents, 89
  detection method, selection, 92
  detection resolution, 90, 90f
    direct immunocytochemistry, 90
    highest detection resolution, 90
    lowest detection resolution, 90
  detection sensitivity, 90, 90f
    high detection sensitivity, 90
    low detection sensitivity, 91
  evaluation of labels and methods, 91f
  experimental design chart, 93–95

fluorescence or enzyme-based labels,
    89–90
  microscopy, 90
    confocal microscopy, 90
    fluorescence microscopy, 90
  planning application method, guidelines, 92
Application methods
  ABC immunocytochemistry
    advantages/disadvantages, 71–73
    avidin–biotin molecules, 68–69
  direct avidin–biotin immunocytochemistry
    advantages/disadvantages, 69–70
  direct immunocytochemistry, 66f
    advantages/disadvantages, 66–67
    bound to 1° antibody, 66
  indirect avidin–biotin
      immunocytochemistry
    advantages/disadvantages, 70–71
    2° antibody binding to 1° antibody, 67
  indirect immunocytochemistry
    advantages/disadvantages, 67–68
Applying fixatives
  dissecting area of interest, 25–26
    orientation of tissue blocks, 26f
  fixating cultures of attached cells,
      problems, 24–25
  fixing suspended cells, 25
  methods
    drop-in fixation, 24
    vascular perfusion, 24
Aqueous embedding media, 106, 183
  Crystal/Mount or Gel/Mount, 106
  Immumount, 106
  Mowiol 40–88 (powder) Sigma, 106
  Prolong Gold, 106
Aqueous mounting media, 106
Aqueous resins
  LR White or Lowicryl K4M, examples,
      183
  postembedding electron microscopic
      immunocytochemistry, use in,
      183–185
Ascites fluid, **12–14**
Autofluorescence, 20, 22, **85–87**
  aldehyde groups of fixatives, 86
  characteristics, 85
  emission spectra, 85
  protocols for, 87
  reduction methods, 85–86
    "aging pigment," 86
    aldehyde groups of fixatives, 86
    emission spectra, 85
    scanning cells and tissues, 86

signal-to-noise (background
    autofluorescence) ratio, 85
spectral imaging, 86
signal-to-noise (background
    autofluorescence) ratio, 85
Avidin, **68**, 69f, 70–72, 74–76
binding sites (high-affinity) for biotin, 69f
Avidin-biotin
molecules, 68–69
reagents, 69f
Avidin-Biotin Complex (ABC), 71–73, 75

B

Background labeling, 14, 102, 103f, 104–105,
    110, 129–130, 137, 157, 170,
    176, 183
Band pass filter set, 129, 137, 142, 143f
Baschong, W., 87
B-cell clone, **10**, 10f, 11–12
Beam splitter, *see* Dichroic mirror
Beisker, W., 87
Bendayan, M., 183
Billinton, N., 85
Biotin, **68–69**, 69f, 70–76
Bivalent epitope binding, 9
Bleed-through, fluorescent, 60, 64, 142–145,
    144f
identification of, 145
micrographs, 145f
Blocking agents, 47, 48f, 49, 51–52, 84,
    154–155, 158, 165, 174, 177
for charged groups
    albumin, 47
    BSA, 47
detergents, 49
effect on antibody penetration, 51–52, 52f
for endogenous antibody
    IgG Fab fragments, 48
for Fc receptors
    normal serum with IgGs, 48
gelatin, 49
glycine, 49
milk (nonfat, freeze-dried), 49
rule for blocking serum, 48
triton X-100, 49
Blocking and permeability
antibody penetration
    blocking agents, effects on, 51–52
    permeabilizing tissue/cells for, 49–51
combined incubation step, 53
nonspecific antibody binding to tissue/cells,
    45–47
    binding sites, 46f
    blocking (agents), 47–49, 48f

Fab-epitope binding, 45
nonspecific binding, 46
'Bloto,' *see* Milk (nonfat, freeze-dried)
Bouin's fixative, **22**
Bovine serum albumen (BSA), **47**–49, 51–52,
    94t, 99, 107–108, 114t, 116, 123t,
    128, 132t, 136–137, 153–155, 157,
    159, 165, 168, 185, 187
Bright field microscope, 64, 139
Buffer solution or 'vehicle'
pH and tonicity, features of, 22–23
    tonicity on cells, effects of, 23f
phosphate buffer, 22–23
pK or pH point, 22, 23f
use in fixation, 22–23
Burry, R. W., 82, 180–181

C

Cauller, L. J., 87
Chromogens, **61**–64, 74, 76, 84–85, 100, 139,
    176, 197t
glucose oxidase, use in development, 62
"Chuck" (metal platform), 33
Clancy, B., 87
Colloidal gold, 182–183, 185
Colocalization, 63–64, 81, 82f, 144
Confocal microscopy, 51, 60, 90, 177
Constant region (IgG), **8**, 8f
Controls, 79–88
    1 antibody controls, 80–83
        recommendations, 83–84
    2 antibody controls, 84–85
    labeling controls, 85–88
        protocols for autofluorescence, 87
        protocols for endogenous
            peroxidase, 88
Coons, A. H., 5, 58
Couchman, J. R., 173
Cross-linking fixation, 19
formaldehyde cross-links groups, 19
glutaraldehyde, 21–22
PLP, 21
Cryostat, 29–**30**, **33**, 37f, 40, 42
protocol, 42–44
sectioning, 33–37
    mounting frozen tissue on chuck,
        34f, 36f
See also Microtome
Cryo-ultramicrotome, 178, 182f, 183, 182f
Cryoultramicrotomy, 183
Cultured cells, 5, 12, 19, 32, 38, 39f, 47, 85,
    87, 122, 178, 181–182
Cupric sulfate reduction, 87
protocol for autofluorescence, 87

D

Denaturing fixation, **18**
Detection controls, *see* Labeling controls
Detection method, 73, 84, 90–92, 102
Detection resolution, 64, **90**, 90f, 91, 91t,
    92–93
    direct immunocytochemistry, 90
    highest/lowest, 90
Detection sensitivity, 62–64, 68, 71–75, 77, 90,
    90, 90f, 91, 91t, 92
    high detection sensitivity, 90
    low detection sensitivity, 91
Detergents
    detergent extraction of membrane
        lipids, 51t
        room temperature, preferred, 51
    ionic/nonionic, 50, 50f
    saponin or digitonin, detergent-like agents
        transient permeabilization of cell
            cultures, use in, 50
3′=3′ diaminobenzidine (DAB; 224 kDa),
        chromogen, 61
Dichroic mirror, 140–141
Diffuse/uniform autofluorescence, 85
Dilution Matrix, 93
    See also Antibody dilution matrix
Direct immunocytochemistry method, **66–67**,
    66f, 91
Drop-in-fixation, **24**, 28

E

Electron microscopic immunocytochemistry,
    175–189, 184f
    choice of method
        pre/post-embedding,
            advantages/disadvantages, 185
    common labels used
        gold and silver, 176
        HRP, 176
    labelling, problems associated, 176
    need for, 176–178
    particulate labeling, 176
    postembedding, 181–185
    pre-embedding (protocol), 175–176,
        178–181, 185–189
    resolution, significance, 175–176
    vs. light microscopic
        immunocytochemistry, 177
Emission filter, 140f, **140–141**, 141f, 142,
    143f, 144
Emission spectra, 56, 85, 141f
Endogenous fluorescence, 85, 86f
Endogenous peroxidase activity, 63, **86–87**

Enzyme activity, 2–4, 61, 85–87
Enzyme amplification methods, 63
Enzyme theory
    detection of low-copy antigens,
        application, 61
    enzyme amplification methods, 63
        multiple 1 antibodies with enzymes,
            problems, 63
    enzyme labeling, 62f
        problem/difficulty, 63
    enzyme substrates
        DAB, 61–62
        HRP, 61
        TMB, 62
    as labels for immunocytochemistry,
        advantage, 61
Epifluorescence, 92, 140
Epifluorescence microscopy, *see* Fluorescence
    microscopy
Epitope, **9–10**
Epitope retrieval/antigen retrieval, 2, 20, **41**,
    42t
Epoxy resin, 176–178, 182–183
Excitation/emission photons, 56–57, 57f, 58
Excitation filter, 140–141, 141f
Experimental design chart, 93–95, 94f
    antibody detection methods, 93
    categorization, 93–95
        1° antibody, 93
        2° antibody and label information, 93
        controls, 95
        incubation solutions, 94t
        microscopy, 95
        sample preparation, 93
    parameters, 93
    reagent selection, 93
    types of information, 93
        Antigen No. 1 and Antigen No. 2, 93

F

Fab binding, 45, 130
Fab-epitope binding, 45–46
F_c receptor, 8, 13, 46f, 47–48, 172, 182
Filter sets
    band pass filter set, 143f
        ideal for multiple fluorophores, 142
    fluorescent microscopy, use in, 140–142
    function of, 141
    long pass filter set, 143f
        ideal for single fluorophores, 142
    parts
        emission filter, 140–141
        excitation filter, 140–141

selection of fluorophores for multiple
    labels, criteria, 142
Fixation, **18**
Fixation theory, 18–19
    fixation of cells, **18**
    good fixation, criteria for, 18
    types of fixation
        cross-linking fixation, 19
        denaturing fixation, 18
    use of fixatives, criteria for, 18
Floating section immunocytochemistry, 40
Flow cytometry, 4, 59–61, 66, 173
    fluorophores and, 59–60, 60f
Fluorescence microscopy, 51, 59–60, 89–90,
    140–142
Fluorescence or enzyme-based labels, 89–90
Fluorescence resonance energy transfer
    (FRET), 58
Fluorescence theory, 56–58
    fluorescence emission loss, causes
        photobleaching, 58
        quenching of emission photons, 57–58
    fluorescent molecules
        excitation or absorption of, 56
        photon emission, 56
        property of, 56
        selection of, importance, 56–57
    fluorescent spectra with excitation and
        emission, 56, 57f
    Alexa Fluor 488 excitation in dark,
        example, 56
    relaxed singlet electron, 56
Fluorescent immunocytochemistry, 4–5, 74,
    79, 86f, 102, 178–179
Fluorescent indirect immunocytochemistry, 91
Fluorescent labeling, generations of, 22,
    52, 56, 58–60, 73, 76, 85–86,
    90, 106
    1st generation 1942
        fluorescein and rhodamine
            fluorophores, 58
    2nd generation 1993
        Cy fluorophores; cyanine dyes, AMCA,
            Texas Red, 58
    3rd generation 1999
        Alexa Fluor fluorophores/
            ATTO/DyLight Fluor, 58
    4th generation 2003
        Quantum dots (Q-dots), 59
Fluorescent microscopy and imaging, 139–149,
    140f
    filter sets, use in, 140–142
        dichroic mirror, key element, 140–141

fluorescence quench and photobleach,
    145–146
fluorescent bleed-through, 142–145
image manipulation, ethics of, 148–149
image parameters, contrast/pixel saturation,
    146–148
Fluorescent molecules (fluorophores), 56
    excitation or absorption of, 56
    photon emission, 56
    property of, 56
    selection of, importance
        quantum yield, factor for, 57
Fluorescent recovery after photobleaching
    (FRAP), 58
Fluorochromes, 56, 58–61, 114–115, 137, 145,
    173
    APC, 60
    choosing, guidelines for, 61
    PE, 59
Fluorophores (fluorescent molecules)
    Alexa Fluor dyes (third-generation),
        146, 147f
    Cy dyes (second-generation), 146
    FITC and rhodamine (first-generation), 146
Formaldehyde (CH_2O)
    chemical structure of, 19f
    cross-links groups, 19
    polymerization of, 20f
    sources of
        formalin/paraformaldehyde, 20
Formalin, 2, **20**–22, 41–42, 47, 49, 86
FRAP, *see* Fluorescent recovery after
    photobleaching (FRAP)
Freezing, methods, 31f, 33f
    dry ice method, 30–31
    plastic molds, use of, 32–33
    plunging the tissue into isopentane, 30
    slow freezing method, 31–32
    strips of aluminum foil, use of, 32
    vitrification, 30
Freezing microtome, 25, 39–41,
    40, 40f
    application, 40
    brain tissue, use in, 40
    used to generate artifacts, 40
Freezing tissue, 32–33
    isopentane, freezing agent, 32
    liquid used
        O.C.T., 32
        TFM, 32
    theory of, 30–32
FRET, *see* Fluorescence resonance energy
    transfer (FRET)

Frozen fixed tissue, 30
Furuya, F. R., 177

G
Gabel, D., 82, 86
G-coupled receptors, 10
Gelatin, 25, 49
GFPs, *see* Green fluorescent proteins (GFPs)
Ghitescu, L., 182
Gilerovitch, H. G., 180, 184f
Glazer, A. N., 59
Glucagon-like Peptide (GLP), 187f
Glucose oxidase, 62
Glutaraldehyde (C5H8O2), **21–22**, 22f, 47, 49,
        85–86, 178, 182–183, 185–186
Glycine, 49, 86
Golgi apparatus, 91, 183f
Good, N. E., 17, 22
"Good's buffers," 22
Green fluorescent proteins (GFPs), 61
Green fluorophore (Alexa Fluor 488), 56, 58,
        60f, 141–142

H
Hainfeld, J. F., 177
Heffer-Lauc, M., 51
Herzenberg, L. A., 59
High detection sensitivity, 64, 68, 75, 90, 92
Histochem Cytochem, J., 51
Histones, 19, 47
Hobot, J. A., 183
Hoffman, G. E., 27, 102
Horseradish peroxidase (HRP), 16, **61**, 62f
        avidin–biotin immunocytochemistry, 91
        intensification methods
                NiDAB, advantages, 62
                oxidizing ionic silver, 62
        substrates, chromogens, 61
Houser, C. R., 84
HRP, *see* Horseradish peroxidase (HRP)
Hsu, S. M., 71
Hybridoma cells, 11–13, 11f
Hydrophobic pen, 37
'Hypertonic' solution, 23–24, 23f
'Hypotonic' solution, 23f, 24

I
IgG antibody isotype
        bivalent epitope binding, advantage, 9
        constant region, 8f
                consists of F_c portion, 8–9
        injection in rabbits
                rabbit anti-mouse IgG antibodies, 9
        long/short protein chains, contents, 9

structure of, 8–9, 8f
subclasses, 8
variable region, 8f
        consists of fraction antigen binding
        (*Fab*) portion, 8–9, 8f
IgG isotypes, 8
Immunoblot (IB)/western blot, 14, 49, **80–81**,
        82–83, 95, 155
Immunocytochemistry, **2**
        animal tissue research, methods, 2
        antibodies, "over fixed"
                epitope retrieval/antigen retrieval
                methods, 2
        antibody labeling methods, 1
        cell analysis, role in, 2, 4
        change in definition, 2
        detergents, use in
                ionic and nonionic, 50, 50f
                nonionic detergents, preferred, 50
        and immunohistochemistry, distinction,
                1–2
        individual studies/population studies,
                comparison of results, 3
        key elements
                access to antigens by antibodies,
                        providance, 45
                binding of antibodies to appropriate
                        sites, 45
        morphological approach
                liver cells, identification of, 3
                protein binding, use of antibodies for, 3
        nonspecific binding, problems/difficulties
                charged groups, localization problem,
                        46–47
                endogenous antibodies, tissue
                        species same as primary antibody
                        species, 47
                incorrect labeling by F_c receptors, 47
        paraformaldehyde, use of, 21
        population study, example, 2–3, 3f, 4
                flow cytometry, 4
                procedure, steps involved, 4
Immunofluorescence (IF), 14
Immunoglobulins (Igs), 8
Immunohistochemistry (IHC), **1**, 14, 21, 41, 79
        use of formalin for clinical or diagnostic
                samples, 21
Immunoprecipitation (IP), 14, 81
Incubation
        solutions, 37–38, 53, 85, 94t, 99t, 100,
                114t, 123t, 132t
        temperature, 100

room temperature over ice temperatures, preferred, 100
Indirect immunocytochemistry, 4, **67–68**, 70–71, 73, 76, 79, 84t, 91, 98f, 99t, 106–107, 114t, 115, 120, 122, 123t, 130, 132t, 168
  single 1 antibody
    with Antigen Ag A, 97, 98f
    with goat anti-mouse 2 antibody, 97, 98f
    list of steps, 106–110
    with mouse anti-Ag A antibody, 97, 98f
Individual studies, 3
Infiltration, 30
Invitrogen (molecular probes), 56, 58–59, 106, 120, 130, 133, 145f
Ionic detergents
  3D protein structures, destuction
    CHAPS, 50
    deoxycholate, 50
    SDS, 50
  solubilizing agents, 50, 50f
Irregular or particulate autofluorescence, 85
Isopentane, freezing agent, **30**, 31f, 32–33, 42–43, 93, 94t, 99t, 107–108
'Isotonic' solution, 23

K
Knight, A. W., 85

L
Labeling controls, 85–88, 86f
  autofluorescence, 85
    aldehyde groups of fixatives, 86
    characteristics, 85
    emission spectra, 85
    signal-to-noise (background autofluorescence) ratio, 85
  black "X" (blocked endogenous activity), 86
  endogenous fluorescence or enzyme, 85
  endogenous peroxidase activity, 86–87
  enzyme activity, 85
  fluorescent labeling, 85
  protocols for autofluorescence, 87
    cupric sulfate reduction, 87
    sodium borohydride reduction, 87
    sudan black reduction, 87
  protocols for endogenous peroxidase, 88
    hydrogen peroxide, 88
    hydrogen peroxide and methanol, 88
    phenylhydrazine, 88
  reducing autofluorescence, methods, 85–86
    "aging pigment," 86

aldehyde groups of fixatives, 86
emission spectra, 85
scanning cells and tissues, 86
signal-to-noise (background autofluorescence) ratio, 85
spectral imaging, 86
spectral confocals or spectral camera, 86
Labels, **56**
  types
    enzymes, 56
    fluorescence, 56
    particulate label, 56
Larsson, L. -I., 5
Long pass filter set, 142, 143f

M
McGhee, J. D., 19
McLean, I. W., 21
Methods of freezing, 31f, 33f
  dry ice method, 30–31
  plastic molds, use of, 32–33
  plunging the tissue into isopentane, 30
  slow freezing method, 31–32
  strips of aluminum foil, use of, 32
  vitrification, 30
Methyl hydrate polymer, 20
Microscopy, 90, 95
  confocal microscopy, 51, 60, 90, 177
  electron microscopy, 21, 22f, 50, 64, 177, 181
  fluorescence microscopy, 51, 59–60, 85, 90, 139–149, 154
Microtome, 25, 29–30, 33, **40**
Microwaves
  application in tissue processing, 40–41
  PELCO BioWave®, example, 41
Milk (nonfat, freeze-dried)
  blocking agent, 49
  western or immunoblots, use in, 49
Monoclonal antibodies, 11
  advantages/disadvantages, 12
  generation process (Kohler and Milstein), 11f
    rabbit monoclonal antibodies, use of, 12
    spleen cells fusion with myeloma cells, 11–12
    supernatant culture media, 12
Mouse myeloma cell line, 11
Multiple antibodies for different species, 111–118
  combining two 1° antibody incubations, 112
  complete final procedure, 115–117

Multiple antibodies for different species (*cont.*)
    1° antibodies, 115–117
    2° antibody, 113
    block and permeabilize, 116
    rinse after 1° antibody, 113
    rinse after 2° antibody, 118
    rinse after block and permeabilize, 116
  designing 2° antibody controls, 113
  experimental design chart, 112–113, 114t
    controls for multiple antibodies
      different species, 115t
    multiple 1° antibodies different species,
      113f
  rules for multiple label experiments,
    113–115
  strategies, two general, 111
    1° antibodies made in different species,
      111
    1° antibodies made in the same species,
      111
Multiple antibodies for same species
  block-between method
    1 and 2 antibody incubation, steps,
      121–122
    2 antibody control, design of,
      124–127
    combining two 1 antibodies, 120–122,
      121f, 120–122
    experimental design chart for, 122

N
Nakane, P. K., 21, 61
Negative control/technique control
  (2 antibody control), 84
Newman, G. R., 86–87, 183
Nickel enhancement or nickel-DAB
  (NiDAB), 62
NiDAB, *see* Nickel enhancement or
  nickel-DAB (NiDAB)
Nonionic detergents
  Triton X-100/Tween 20, 50
Nonspecific binding, 46, 84
  binding sites
    charged groups, 46f
    correct antigen, 46f
    endogenous antibodies, 46f
    F_c receptors, 46f
  problems for immunocytochemistry
    examples, 46f
  sources of, 46f
Normal serum, 47, 48
  vs. cloudy serum, 48
North, A. J., 149

O
Oligodendrocyte, 158–159
Optimum cutting temperature (O.C.T.), 32
Osmoles, 24

P
Papain (enzyme), 9, 8f, 122
Paraffin, 1–2, 20–21, 29, 41–42
Paraformaldehyde, 20, 20f
  animal research experiments, use
    in, 20
  vs. formalin, advantage as fixative, 20
Particulate labeling, 176
PBS, *see* Phosphate buffered saline (PBS)
PE, *see* Phycoerythrin (PE)
Pearson, H., 149
PELCO BioWave®, 41
Perfusion
  equipment, 28
  procedure, 27–28
Periodate-lysine-paraformaldehyde
  (PLP), 21
Permanox, 38–39
Permeabilization of cells, 45
Permount (resin-based organic medium),
  39, 62
Peschke, P., 62
PH buffering, 23f
Phenylhydrazine, 87–88
Phosphate buffered saline (PBS), 26, 107
Photobleaching, **58**
  approaches
    antifade agents, use of, 146
    exposure of fluorophores to excitation
      light, reducing, 146
    fluorophores dyes, use of, 146
  basis for FRAP, 58
  FITC, example, 58
Phycoerythrin (PE), 59, 60f
Pierce, G. B., 58, 61
PLP, *see* Periodate-lysine-paraformaldehyde
  (PLP)
Polak, J. M., 5
Polyclonal antibody, advantages/disadvantages,
  9–11, 10f, 13
  chicken polyclonal antibodies, 10
  serum from immunized rabbit with B-cell
    clones, 10, 10f
Population studies, 3
Postembedding electron microscopic
  immunocytochemistry, 176–178,
    181–185, 182f
  aqueous resins, use of, 183

labeling, use of colloidal gold, 183
limitation of, 183–185
procedure, 177t
solutions
   NPG silver enhancement solution and
     silver lactate, 188
   solutions made on first day of the
     experiment, 187–188
   stock solutions prepared and stored,
     186–187
   test strip, 189
Pre-embedding electron microscopic
   immunocytochemistry, 178–181,
   179f
labeling, use of colloidal gold, 182
procedure, 177t
   1 antibody incubation, 178–179
   2 antibody incubation, 179–180
   permeabilization and blocking step,
     178–179
   rinses, 178–179
   silver enhancement of small gold, 180,
     180f
   standard electron microscopy, 181
Primary (1) antibody, 15, 66
Protein–DNA cross-linking, 19
Protocols for autofluorescence, 87
   cupric sulfate reduction, 87
   sodium borohydride reduction, 87
   sudan black reduction, 87
Protocols for endogenous peroxidase, 88
   hydrogen peroxide, 88
   hydrogen peroxide and methanol, 88
   phenylhydrazine, 88
Purification of antibodies
   affinity purification, 13
   IgG purification, 14
   separation from serum proteins, 13

Q

Quantum dots (Q-dots), 59
   in labeling, advantage/disadvantage, 59
Quantum yield, fluorophore, 56
Quenching, 57–58, 145
   basis for FRET, 58

R

Rabbit anti-tubulin antibody, 13
Red fluorophore (Alexa Fluor 555), 142
Relaxed singlet electron, 56, 57f
Renshaw, S., 5
Riggs, J. L., 58
Rossner, M., 149

S

Sabatini, D. D., 21
Sample preparation/fixation
   applying fixatives
     dissecting area of interest, 25–26
   buffer solution or vehicle solution, 22–24
     pH and tonicity, features of, 22–23
   chemical fixatives, 19–22
     acrolein, 21
     binding/retention of DNA, 19
     formaldehyde, chemical structure of,
       19f
     formalin/paraformaldehyde, 20
     glutaraldehyde, 21–22, 22f
     PLP, 21
     time of fixation, 19
   fixation theory, 18–19
   procedure
     drop-in-fixation, 28
     perfusion equipment, 28
     perfusion procedure, 27–28
   protocol – fixation
     paraformaldehyde fixative, components
       for, 26–27
Saponin or digitonin, 178–179
Schmiedeberg, L., 19
Schnell, S. A., 86–87
Shu, S. Y., 62, 146
Signal-to-noise (background autofluorescence)
   ratio, 85
Silver enhancement process, 177–178
Single antibody procedure
   2 antibody controls, 102–103
   antibody dilution matrix, 102
   antibody dilutions, 100–101
     calculation of, 101
     optinal dilution determination, critical
       step, 101
     ratios, 101
   experimental design chart, 98, 99t
   final procedure
     immunocytochemistry experiment for
       Ag A, steps, 107–110
     indirect immunocytochemistry
       experiment, steps, 106–107
   incubation conditions
     agitation of samples, 100
     incubation temperature, 100
     incubation time, 100
     penetration of antibodies, 98–100
   mounting media, 105–106
     antibleaching agents, use, 106
     aqueous mounting medium, 106

Single antibody procedure (*cont.*)
    refractive index, feature of, 106
    rinses
        after 1 antibody before 2 antibody,
          104f
        after 2 antibody before mounting a
          coverslip, 105f
Snap freezing, 32
Specificity control methods, 80, 82f
    immunoblot (western blot), 80–81
    immunocytochemical localization,
        comparison of, 81
    tissue sections, comparibility, 80
Spectral imaging, 86
Standard electron microscopy, 181
Staphylococcus aureus, 13
Sternberger, L. A., 5
Stirling, J. W., 176
Stokes shift, **56**, 57f, 59, 60f
Streptavidin, 68
Stryer, I., 59
Sudan black, reduction of autofluorescence, 87
SUPERFROST® PLUS slides, 35
Supernatant, **12**–14
Swaab, D. F., 82

T
Tetramethylbenidine (TMB), 62
Theory of freezing tissue, 30–32
    methods of freezing, 31f, 33f
        dry ice method, 30–31
        plastic molds, use of, 32–33
        plunging the tissue into isopentane, 30
        slow freezing method, 31–32
        strips of aluminum foil, use of, 32
        vitrification, 30
Tissue freezing medium (TFM), 32
Tissue processing, 37–39
    chamber slides for cell cultures, use of,
        38–39, 39f
    incubation of cultures on coverslips, 39f
    incubation of tissue sections, 38f
Tissue sectioning
    cryostat protocol, 42–44
    cryostat sectioning, 33–37
    embedding tissue by freezing
        cryostat sectioning, 30
        freezing the tissue, 30
        infiltration, 30
    embedding tissue with paraffin
        epitope/antigen retrieval (processing
          methods), 41
        fixation and embedding methods,
          comparison of, 42t

freezing tissue, 32–33
fresh frozen tissue, 41
theory of freezing tissue, 30–32
tissue processing, 37–39
vibratome, freezing microtome, and
    microwave, 39–41
TMB, *see* Tetramethylbenidine (TMB)
Transcardiac perfusion, *see* Vascular perfusion
Transmembrane proteins, 49
    role in permeabilization, 49
    scaffold proteins, example, 49
Triton X-100, 49, 178
Troubleshooting, immunocytochemistry
    bad antibodies, 173–174
    bad 1 antibodies, 173–174
    bad 2 antibodies, 174
    high background staining, 170–172
    method: case 1, 153–156, 155f
        distributions of ribosome proteins in
          mouse spinal cord, 153
        hypothesis generation, 154
        investigation, 154–156
        problem defined, 153–154
        solution, 155f
    method: case 2, 157f
        1 antibody for a synaptic vesicle
          protein p38, 156
        hypothesis generation, 158
        investigation, 158
        problem defined, 156–157
    method: case 3, 158–163, 160f, 162f–163f
        distribution in gray mater of spinal cord
          of glial cell, examining, 158–163
        hypothesis generation, 159–161
        investigation, 161
        problem defined, 159
    method: case 4, 164f
        hypothesis generation, 165
        investigation, 165–167
        neuronal synapses in white matter of
          mouse spinal cord, investigation,
          164
        problem defined, 164–165
    method: case 5, 167f, 169f, 171f, 172f
        hypothesis generation, 168
        investigation, 169–170
        problem defined, 168
    procedural errors, 152
    sources of problems, 151–152
    unique to multiple primary antibodies, 173
Two 1 antibody (same species), final
    procedure
    block-between experiment, first approach

block and permeabilize, 127
block antibodies in first set, 128
cell culture preparation, 127
fixing culture, 127
incubation, 128
microscope examining, 129
mount coverslip, 129
result evaluation, 129–130
rinsing, 128
with zenon, second approach, 131f, 132t
blocking and permeabilization, 136
cell culture preparation, 135
examining cultures using microscope,
137
experimental design chart for, 130–133
fixation with 4% paraformaldehyde,
136
fixing culture, 135
incubation with labeled antibody(ies),
136
mount coverslip, 137
no first 1 labeling, 134
no second 1 labeling, 135
result evaluation, 137
rinsing, 136
zenon reagents, preparation of, 136
See also Multiple antibodies for same
species
Tyramide signal amplification (TSA),
73–75

U
Upright microscope, 139

V
Van der Loos, C. M., 63, 86–87
Van Noorden, S., 5
Variable region (IgG), 8–9, 8f
Varndell, I. M., 184f
Varshavsky, A., 19
Vascular perfusion, 24
Vectashield ®, 106
'Vehicle'
pH and tonicity, features of, 22–23
tonicity on cells, effects of, 23f
phosphate buffer, 22–23
pK or pH point, 22, 23f
use in fixation, 22–23
Vibratome, 25, **39–40**
Vitamin B$_7$, see Biotin
Vitrification, 30
Von Hippel, P. H., 19

W
Western blot (WB), 14, 80, 174, 181
Whole serum, 13–14
Wolf, W. P., 82

Y
Yamada, K. M., 149

Z
Zamboni's fixative, 22
Zenker's fixative, 22
Zenon mouse IgG labeling reagent, 133
Zenon®, 120, 130–137
reagent preparation, 136
Zollinger, M., 183

9 781441 913081